KEEPING MAJOR NAVAL SHIP ACQUISITIONS ON COURSE

Key Considerations for Managing Australia's SEA 5000 Future Frigate Program

JOHN F. **SCHANK** MARK V. **ARENA** KRISTY N. **KAMARCK** GORDON T. **LEE**

JOHN **BIRKLER** ROBERT E. **MURPHY** ROGER **LOUGH**

For more information on this publication, visit www.rand.org/t/RR767

Library of Congress Cataloging-in-Publication Data is available for this publication
ISBN: 978-0-8330-8818-5

Published by the RAND Corporation, Santa Monica, Calif.

Cover design by Peter Soriano

Preface

The Commonwealth of Australia is pursuing options and opportunities to develop and construct the next generation of naval surface combatants, termed the Future Frigate. The SEA 5000 program was recently established to help decisionmakers understand the implications of various acquisition options for the new combatant program. Currently, SEA 5000 is in pre-Phase 1, focusing on setting up the essential design and construction elements that a major acquisition program needs. During this stage, it is important for the program to have a road map of various paths and steps that it could take in designing, constructing, testing, and supporting a naval ship. The road map should portray major milestones and decision points and the potential implications of those decisions as the program progresses. It should show options for the various potential paths through the life cycle of the program and the implications of following a particular path or paths, including the effect on other options and paths.

RAND is providing materiel studies and analysis activities to support the SEA 5000 Future Frigate Design and Construction Work Package 3. The Work Package is intended to inform the continued refinement of program life-cycle options by providing a program overview of the Naval Shipbuilding Capability Life Cycle. The desired output of the Work Package is a definition of the acquisition options available for the Future Frigate, an overview of a naval ship acquisition program (with particular attention to important considerations SEA 5000 must make early in the program), and an identification of the internal and external factors that can influence a major ship acquisition program. This report provides that information, with the presumption that the reader has a general understanding of defense acquisition. Our work was sponsored by Australia's SEA 5000 Future Frigate program office.

This research was conducted within the Acquisition and Technology Policy Center of the RAND National Security Research Division (NSRD). NSRD conducts research and analysis on defense and national security topics for the U.S. and allied defense, foreign policy, homeland security, and intelligence communities and foundations and other nongovernmental organizations that support defense and national security analysis. For more information on the Acquisition and Technology Policy Center, see http://www.rand.org/nsrd/ndri/centers/atp.html or contact the director (contact information is provided on the web page).

Comments or questions on this report should be addressed to the project leader, John Schank, at John_Schank@rand.org or 703.413.1100 Ext. 5304.

Contents

Preface ... iii
Figures ... xi
Tables ... xiii
Summary ... xv
Acknowledgments ... xxv
Abbreviations ... xxvii

CHAPTER ONE
Introduction ... 1
Background ... 1
Design and Build Options ... 2
Ship Acquisition Process ... 2
Factors That Contribute to Program Success ... 3
Organization of the Report ... 4

CHAPTER TWO
Establishing and Supporting a Program ... 7
Be an Intelligent and Informed Partner in the Shipbuilding Enterprise ... 8
Delineate Roles, Responsibilities, and Decisionmaking Authority ... 10
Establish and Support a Program Office ... 13
Involve All Appropriate Organizations ... 14
Obtain and Sustain Political Support ... 15
Develop Knowledgeable and Experienced Managers ... 15
Develop Realistic Cost and Schedule Estimates ... 17
Understand the Effects of External Factors Beyond the Program's Control ... 20

Government Funding Decisions 21
Industrial Base Issues and Market Conditions 23
Economic Matters 24
Take a Long-Term, Strategic View 24
Key Points in Establishing and Supporting a Program 26

CHAPTER THREE
SEA 5000 Acquisition Options 27
Pure MOTS Option 30
New Design, Build in Australia Option 32
Evolved MOTS Options 33
Important Considerations with Different Acquisition Options 35
Operational Requirements 35
Schedule Constraints 35
Program Costs 36
Technical Risks 37
Desire for Competition 37
Industrial Base Policies 38
In-Service Support and Modernization 38
Summary Comments 38
Important Initial Program Questions 39

CHAPTER FOUR
Overview of a Naval Shipbuilding Program 43
How Are Ships Different from Other Weapon System Acquisitions? 43
Ship Program Phases 45
Design-Build Sequence 48
Sequential Design-Build 48
Concurrent Design-Build 51
Program Oversight: Major Decision Milestones 55
General Timeline of the Process 60
Ship Construction and Testing Milestones 61

CHAPTER FIVE
Acquisition and Contracting Strategy 65
Government Interaction with Industry 65

Different Types of Contracts ... 68
Allocate Contract Risk Appropriately Between Government and Industry ... 71
Obtain Needed Intellectual Property and Technical Data Rights ... 73
Develop Processes for Managing Contract Changes ... 76
Establish an Agreed upon Tracking Mechanism and Payment Schedule ... 77
Key Points with Acquisition and Contracting Strategies ... 79

CHAPTER SIX
The Solutions Analysis Phase ... 81
Set Operational Requirements ... 82
Identify Important Considerations ... 84
Analysis of Alternatives ... 84
Industrial Base Resource Assessment ... 85
Future Upgradability and Margins ... 85
Computing Architectures ... 89
Involve All Appropriate Organizations in Setting Operational Requirements ... 90
Remember That the Ship Is an Integration of Various Systems ... 91
Clearly State Requirements ... 92
Develop a Test Plan for Requirements ... 93
Keep Requirements Stable ... 94
Support Requirements Through Relevant Technical, Safety, and Classification Standards ... 95
Key Points for the Solutions Phase ... 96

CHAPTER SEVEN
Design Activities ... 99
Concept Refinement/Design ... 99
Definition and Activities ... 99
Important Considerations ... 100
Preliminary Design ... 103
Definition and Activities ... 103
Important Considerations ... 103

Contract Design 104
Definition and Activities 104
Important Considerations 105
Detailed Design 106
Definition and Activities 106
Important Considerations 106
Required Resources 107
Personnel—Skills 107
Product Models 109
Private Sector Versus Government 109
Design-Related Factors the Program Can Control 110
Ensure That the Selected Design Organization Understands the Concept of Operations and Build Environment 111
Specify Adequate Design Margins and Manage Them During the Design and Build Program 112
Include in the Design the Capability to Remove and Replace Equipment 113
Understand the Technical and Integration Risks 114
Develop an Integrated Master Plan for Design and Build 116
Consider Potential Problems with Foreign Suppliers 116
Key Points for the Design Phase 117

CHAPTER EIGHT
Manufacturing and Build 119
Definition and Activities 119
Required Resources 121
Personnel—Skills 121
Facilities 122
Establishing a Supplier Base 124
Important Considerations 126
Design Maturity at the Start of Construction 126
Advanced Outfitting 127
Outsourcing 130
Multiple Shipyards 132
Gap Between First and Second Hull 135
GFE Versus CFE Decisions 135

Tracking Progress ... 136
Oversight at Construction Shipyards ... 138
Build-Related Factors the Program Can Control ... 140
Avoid Concurrency in Design and Production ... 140
Develop Effective Coordination and Quality-Control Processes to Support a Distributed Build Strategy ... 141
Ensure That Sufficient Oversight Exists at the Construction Shipyards ... 142
Develop a Management System to Track Progress During the Design and Build Process ... 142
Plan for Operational Testing ... 143
Key Points for the Manufacturing Phase ... 144

CHAPTER NINE
Test and Trials ... 145
Definition and Activities ... 145
Important Considerations ... 148
Post-Trial Availability ... 148
Test Planning ... 148

CHAPTER TEN
Operations and Support ... 151
Begin Integrated Logistics Support Planning Early in the Program ... 151
Maintain Adequate Funding to Develop and Execute the ILS Plan ... 152
Account for Maintenance and Modernization ... 153
Account for Crew Training and Transition ... 154
Consider ILS from a Navy-Wide Rather Than a Program Perspective ... 155
Plan for Technology Advancements and Obsolescence Management ... 156
Key Points in Operations and Support ... 157

CHAPTER ELEVEN
Summary Comments ... 159
Be an Intelligent and Informed Partner in the Acquisition Process ... 160
Involve the Appropriate Organizations Early and Often ... 163
Clearly Assign Roles, Responsibilities, and Risks ... 163
Understand the Cost and Schedule Implications of Options ... 163

Clearly State Requirements 164
Understand and Obtain Required Intellectual Property Rights 165
Strive for Program Stability 165
Start Construction Only When Designs Are Largely Finished 166
Develop an Integrated Logistics Support Plan Early 166
Have a Strategic Perspective 167
Address Critical Near-Term Questions Facing the SEA 5000 Program 167
What Are the Operational and Performance Requirements for the Future Frigate? 168
Which Solution Is the Most Cost-Effective (i.e., MOTS, Evolved MOTS, New Design)? 168
How Should Government Engage with Industry? 168
What Are the Technical Requirements and Risks? 168
How Will the Program Office Monitor the Program? 168
How Will the Class Be Supported Through Its Life? 169

APPENDIXES
A. Overview of Recent Shipbuilding Programs in Australia, the United States, and the United Kingdom 171
B. Technology Readiness Levels 181
C. Best Practices in Scheduling 183

Bibliography 187

Figures

4.1. Sequential Design-Build Process 49
4.2. The Concurrent Design Process with IPPD 54
4.3. Shipbuilding and Acquisition Phases, Decision and Requirements Milestones, and Australian Passes for First Hull 58
5.1. Spectrum of Contract Types and Cost Risk 72
8.1. Vendor Qualification Process 124
9.1. Notional Major Milestones of Test and Trials During Construction 148

Tables

S.1. Key Differences Between Pure MOTS, Evolved MOTS, and New Design Acquisition Options xix
3.1. Summary of Acquisition Options........ 40
4.1. Comparison of Technical Characteristics of Three Programs........ 44
4.2. Advantages and Disadvantages of a Sequential Design Process........ 51
4.3. Advantages and Disadvantages of Concurrent Design Process That Includes IPPD Processes........ 55
4.4. Notional Durations of Acquisition Phases for Naval Ships.... 60
4.5. Comparison of Select Ship Program Characteristics 62
5.1. Example Acquisition Strategies and Organization in Lead Role, by Design-Build Phase........ 68
5.2. Contract Types and Uses in Major Acquisition Projects........ 70
6.1. Notional Acquisition Weight Margins (As a Percentage of Light Ship Displacement)........ 87
6.2. Notional Acquisition KG Margins (As a Percentage of KG in Light Ship Displacement)........ 88
6.3. Service Life Allowances for Weight and KG at Delivery 89
7.1. Technical Skills in Warship Design 108
7.2. Procurement Cost Comparisons........ 113
8.1. Manufacturing Skills in Warship Construction 122
8.2. Facilities Use During Ship Production........ 123
8.3. Shipbuilder Use of Progress Metrics at Various Shipbuilding Phases 137
11.1 Key Differences Between Pure MOTS, Evolved MOTS, and New Design Acquisition Options 161

A.1. Recent Shipbuilding Programs Reviewed for This Study 171
B.1. Technology Readiness Levels Definitions 181
C.1. GAO Scheduling Best Practices 183

Summary

The Royal Australian Navy (RAN) operates eight *Anzac*-class frigates.[1] These warships began to enter service in the mid-1990s and are scheduled to begin inactivation in the mid-2020s. The RAN is transitioning to a future surface combatant force and has identified the need for increased surface combatant capabilities with an emphasis on anti-submarine warfare. This increased capability is planned to be introduced by the SEA 5000 Future Frigate program, and as the ships are introduced, the *Anzac* class with be withdrawn from service.

As with any large, complex military acquisition program, SEA 5000 program managers need to understand various paths and steps that could be taken in designing, building, testing, and supporting a complex naval ship and the implications of the various decisions that might be made along those different paths. Faced with this need for greater insight, in early 2014 the RAND Corporation was commissioned to help explore three acquisition policy questions:

- What ship design and build options are available for the Future Frigate and what are the implications of choosing among the various options?
- What are the various phases, options, and decisions in a naval ship acquisition program?
- What are the important aspects that can contribute to the success of a program?

[1] The Royal New Zealand Navy also operates two *Anzac* frigates.

Three Design and Build Options

RAND examined the acquisition and operational implications posed by three broad categories of options to design and build the Future Frigate that the SEA 5000 program is considering:

- Military Off-the-Shelf (MOTS): Using this option, the RAN would acquire an existing ship design and configuration from an Australian or foreign ship producer or producers; the RAN would make only minor modifications to the ship's design.
- Evolved MOTS: As with the MOTS option, with this option the RAN would acquire an existing ship design and configuration from an Australian or foreign ship producer or producers; however, the RAN would make more significant modifications to the ship's design to reflect evolving, Australia-specific requirements.
- New Design: This option would entail designing a new ship from scratch; sometimes called a clean sheet design, this option would be designed and tailor-made to evolving RAN specifications and requirements.

Each option would entail different risks and implications for the acquisition process and strategy. The pure MOTS solution (which most likely would be built outside Australia) probably would entail the least design and cost risk, given that there would most likely be an experienced builder and warm (i.e., active) supplier base. Evolved MOTS options would entail more design and build risk, which would increase as the ship's design diverged from the baseline design. A new design would present the most acquisition risk, as everything must start from a clean sheet. However, the operational risks would be reversed to some degree. Assuming that each option performs to specification, the clean design specifications could be tailored to the specific needs of the RAN. The pure MOTS option is a fixed design, so the RAN would have to adapt its operations to the ship and may not get every feature that it desires to meet operational requirements.

In this report, we discuss lessons learned as they apply to different phases of a shipbuilding program and attempt to highlight which lessons are most applicable to the acquisition strategy selected by the RAN for the design and build of the Future Frigate. Many of these lessons will apply regardless of the selected strategy. In addition, some overarching lessons can be applied across all phases of the program, from initial requirements development to life-cycle support. We also summarize the key overarching lessons learned to help guide program managers' planning and decisionmaking.

Naval Ship Program Phases

Military ship programs go through eight phases during their life cycles.

1. Solutions Analysis: In this phase, government and industry conduct a broad exploration of materiel solutions that may meet the military's needs. They explore multiple alternative solutions to understand the cost-effectiveness and affordability of various system choices.[2] At the end of this phase, the government selects a single concept to refine in the next phase.
2. Concept Design: During this phase, various military requirements are traded against ship size and cost. The output of the phase should be a conceptual design and preliminary weight for the proposed ship, along with an assessment of major risks.
3. Preliminary Design: This phase fully defines the ship characteristics (major components, hull form, technologies), establishes system architectures, and establishes a detailed cost baseline using the conceptual design produced from the prior phase. Among other things, it produces a general arrangement drawing, system diagrams, and a list of major equipment.
4. Contract Design: This phase traditionally hones the technical and contractual definition of the ship design (including ship specifications and drawings) to a level of detail sufficient for shipbuilders to make a sound estimate of the detailed design

[2] By *system* in this context, we mean some major element of the ship, such as propulsion plant, sensors, a weapon, or aviation support equipment.

and construction cost and schedule. New technology developments are typically initiated during this phase.

5. Detailed Design and Construction: During this phase, the ship design is fully defined in terms of product models, construction drawings, and procurement specifications for material, equipment, and systems. Construction begins as design products are finished. Logistic support plans and crew training materials are also developed during this phase.
6. Test and Trials: In this phase, all the operational aspects of the ship are checked. Test and trials are much more extensive for the first hull to prove the design. Once they are finished, the ship returns for a brief refit and repair period to address any problems and to upgrade certain systems.
7. Operations and Support: In this phase, the ship conducts operations and requires support. The program office still has a role in planning and executing maintenance as well as handling improvements and upgrades.
8. Retirement/Disposal: At the end of its useful life, a ship is sold or scrapped.

How do these phases mesh with the three design and build options? Some of the key differences between the design and build options and the life-cycle phases are shown in Table S.1. For a pure MOTS or evolved MOTS acquisition, the solution analysis phase identifies and evaluates various existing designs and chooses one for further exploration. However, in the case of a pure MOTS acquisition, the next three phases—concept, preliminary, and contract design—can be skipped or carried out quickly, given that the original design is not being changed, whereas an evolved MOTS acquisition will need to go through those phases more thoroughly, given that the design is being changed from the original. An acquisition using a new design will need to progress through the three phases completely. All acquisitions go through the last four phases—detailed design, test and trials, operations and support, and retirement—regardless of whether a MOTS, evolved MOTS, or new design is chosen.

Table S.1
Key Differences Between Pure MOTS, Evolved MOTS, and New Design Acquisition Options

Pure MOTS Option	Evolved MOTS Option	New Design Option
	Solutions Analysis Phase[a]	
Choice between specific designs	Choice between specific design and level of modification	Choice between level of performance
	Design Phase (Concept, Preliminary, Contract, and Detailed)	
Requirements are selected by parent Navy, not RAN	Clear requirements definition important to define level of modifications needed	Clear requirements definition critical
Can begin design process at contract design stage	Will need to progress through all design phases	Will need to progress through all design phases
Detailed design largely complete	Design periods may be short if level of modifications is minimal	Greatest design risk (cost and schedule)
Construction design and instructions depend on build strategy (domestic or foreign)	Moderate design risk (cost and schedule)	Margins can be flexible
Minimal design risk (cost and schedule)	Margins are predetermined and not adjustable	
Margins are predetermined and not adjustable		
	Construction Phase	
Foreign build can leverage existing manufacturing efficiencies	Foreign build can leverage existing manufacturing efficiencies	Will start from no prior experience
	Operations and Support Phase	
Issues on intellectual property (IP) rights and ability to modify design	Issues on IP rights and ability to modify design	Need to design for future updates and modifications
Can potentially leverage an existing parts supply base	Can potentially leverage an existing parts supply base	Least risk that the design will not satisfy RAN's needs (if requirements defined early)
Greatest risk that design will not satisfy RAN's needs	Moderate risk that design will not satisfy RAN's needs	

Table S.1—Continued

Pure MOTS Option	Evolved MOTS Option	New Design Option
	Cross-Cutting Issues	
Acquisition strategy will be limited to a single designer	Acquisition strategy will be limited to a single designer	Many acquisition strategy options possible
Alignment between designer and builder challenging if domestic build option selected	Alignment between designer and builder challenging if domestic build option selected	Classification society open choice
Already chosen classification society (or pay for redesign)	Already chosen classification society (or pay for redesign)	Least IP challenges
Greatest IP challenges	Moderate IP challenges	

[a] Multiple options could be considered during the solutions analysis phase. However at the end of this phase, a single option is taken forward, typically.

Important Aspects That Can Contribute to Program Success

The RAND team highlighted a range of lessons that it believes are most applicable to the RAN's prospective design and build approach to the Future Frigate. Many of the lessons will apply regardless of the selected strategy, and some can be applied across all phases of the program.

Be an Intelligent and Informed Partner in the Acquisition Process

As with any buyer of a major item, the Defence Materiel Organisation (DMO) and the RAN must have knowledge and expertise in what capabilities are desired from the new ship, the current and future technologies that can affect the ship's performance, and the costs, schedules, and risks of adopting different acquisition strategies.

Involve the Appropriate Organizations Early and Often

To develop that knowledge base, it is important to involve technical experts, industry, operators, and maintainers in each phase of the program. In early phases, this varied expertise can help program managers understand both the technical and manufacturing feasibility of achieving speed, weight, or other performance capabilities. It can also help to build a better understanding of the cost trade-offs for certain capabilities or different acquisition strategies. Later, it can help program managers understand how design decisions will affect the ease of maintaining, upgrading, or replacing equipment over the life cycle of the ship.

Clearly Assign Roles, Responsibilities, and Risks

Programs are more successful when there is a clear understanding of the roles, responsibilities, and risk sharing between government and industry, and they should be clearly defined in the contract design phase to prevent future disputes. The government and the contractor should be responsible for cost and schedule risk in the areas under their respective control. In addition, IP rights and ownership of technical data should be negotiated and assigned early in the program.

Understand the Cost and Schedule Implications of Options

Realistic cost and schedule estimates are needed throughout the program. Cost estimates must be based on through-life costs for the fleet of ships and include cost elements ranging from design and development to operations and support to the deactivation and disposal of the ships. Well-informed cost and schedule estimates are especially important when a program begins and decisions are faced on which acquisition option provides the best value for money.

Clearly State Requirements

The DMO and RAN must determine the capabilities desired from the Future Frigate. Operational requirements define the ship's missions and the operational effectiveness in accomplishing those missions. Requirements also include how the ship will operate (i.e., a concept of operations or operational concept document) and be supported during its operational life. These requirements should be expressed in terms of what is desired, not how to specifically accomplish the objectives, and defined in ways that are agreed to by the DMO, RAN, and the ship designer and builder. In addition, the government should strive to avoid any changes to requirements that may affect cost and schedule. The DMO and RAN must also define how the ship will be tested.

Understand and Obtain Required Intellectual Property Rights

When negotiating the contract for a MOTS or evolved MOTS design, it is important to obtain IP rights and ensure that there are no legal barriers to the export of documentation or materials from participating foreign suppliers. These rights are especially important for properly modernizing and maintaining the ships during their operational lives.

Strive for Program Stability

Shipbuilding programs can take more than a decade from initial concept development to full production. Over that time, there will undoubtedly be changes to the external landscape including new technological developments, industrial base developments, and shifts in national strategic or budgetary priorities. Maintaining program stability in the face of these developments requires effective mechanisms to cope with change and to manage stakeholders. It is important to account for change-management processes in the contracting phase and to develop methodologies to assess the cost and effect of changes when they do occur.

Start Construction Only When Designs Are Largely Finished

Starting construction on the lead ship before designs are virtually finalized can lead to costly rework and schedule delays. In some situations, gaps in industrial base demands may favor beginning ship construction when detailed production designs are still evolving. But often it is more cost-effective to delay construction rather than risk the rework and changes that result from design immaturity. A general rule of thumb suggests that 80 percent or more of the detailed design drawings should be complete when construction begins.

Develop an Integrated Logistics Support Plan Early

A robust integrated logistics support (ILS) plan depends on a clear concept of how the Future Frigate will operate and be maintained and must be addressed early in the program and be continuously refined. It should specify what maintenance and updates are required at different points during the Future Frigate's operational life and who will accomplish the required maintenance and modernizations.

Have a Strategic Perspective

The Future Frigate is only one of the RAN's strategic assets. When deciding on capabilities and requirements and selecting an acquisition strategy, it is important to consider how this platform will complement and integrate with the capabilities of other platforms. The Future Frigate program must be viewed in light of overall national objectives for the naval shipbuilding industrial base. It also is important to consider how the existing maintenance, training, and support infrastructure can be leveraged to reap cost savings across the fleet. When possible, program managers should try to strive for commonality in parts, tools, and materials.

Critical Near-Term Questions Facing the SEA 5000 Program

The SEA 5000 program needs to address at least six important, near-term questions that will shape the program for decades to come; these are listed below. Understanding the timing and importance of its decisions on these questions is one key to a successful program.

- What are the operational and performance requirements for the future frigate?
- Which solution is the most cost-effective (MOTS, evolved MOTS, new design)?
- How will engagement with industry be managed?
- What are the technical requirements and risks?
- How will the program office monitor the program?
- What are the earned value management requirements?
- How will the class be supported throughout its life?

Acknowledgments

We very much appreciate the support provided by Glenn Alcock of the SEA 5000 program. He helped us understand the Australian defense acquisition system and the differences between the methods used in the United States and those of other countries. In composing this document, we drew on the knowledge and expertise of many of our RAND colleagues. We have always learned from our interactions with Irv Blickstein, Charles Nemfakos, Paul DeLuca, Roland Yardley, Robert Button, and many others. We appreciate the thorough reviews of an earlier draft of the report by Hans Pung and Michael Hammes. Their insights, comments, and suggestions greatly improved the overall report. Of course, any errors of omission or commission are ours alone.

Abbreviations

ACA	aircraft carrier alliance
AO	afloat outfitting
AoA	analysis of alternatives
APUC	average procurement unit cost
AUS	Australia
AWD	Air Warfare Destroyer
BCWP	budgeted cost of work performed
BCWS	budgeted cost of work scheduled
BY	base year
C4I	command, control, communications, computers, and intelligence
C4ISR	command, control, communications, computers, intelligence, surveillance, and reconnaissance
CAD/CAM	computer aided design/computer aided manufacture
CFE	contractor furnished equipment
CFO	contractor fitting out

COATS	command-and-control off-hull assembly and test site
CONOPS	concept of operations
CPFF	cost plus fixed fee
CPIF	cost plus incentive fee
CVF	*Queen Elizabeth* aircraft carrier; also QEC
DMO	Defence Materiel Organisation
DMSMS	diminishing manufacturing sources and material shortages
DoD	Department of Defence (Australia)
EMI	electro-magnetic interference
EPCM	engineer, procure, and construction management
ETP	Early Test Plan
EVM	earned value management
FA	final assembly
FP	firm price
GAO	Government Accountability Office (United States)
GFE	government furnished equipment
GFE/M	government furnished equipment/material
GM	metacentric height
HM&E	hull, mechanical, and electrical
HVAC	heating, ventilation, and air conditioning
ILS	integrated logistics support
IP	intellectual property
IPPD	Integrated Product and Process Development

IT	information technology
KG	height of the center of gravity above the keel
KPP	key performance parameter
KSA	key system attribute
LCS	Littoral Combat Ship
LHD	Landing Helicopter Dock
LRIP	low rate initial production
MOD	Ministry of Defence (United Kingdom)
MOTS	military off-the-shelf
MOE	measures of effectiveness
MOP	measures of performance
NAO	National Audit Office (United Kingdom)
NAVSEA	Naval Sea Systems Command (United States)
NVR	Naval Vessel Rules
O&M	operations and maintenance
PAUC	program acquisition unit cost
PSA	Post Shakedown Availability
QEC	*Queen Elizabeth* aircraft carrier; also CVF
R&D	research and development
RAN	Royal Australian Navy
RDT&E	research, development, test, and evaluation
ROM	rough order of magnitude
SLA	service life allowance
SUPSHIP	Supervisor of Shipbuilding

T&E	test and evaluation
TCD	Test Concept Document
TCI	target-cost incentive
TEMP	Test and Evaluation Master Plan
TRL	technology readiness level
TY	then year
UAV	unmanned aerial vehicle
VP	variable price

CHAPTER ONE

Introduction

Background

The Royal Australian Navy (RAN) operates eight *Anzac*-class frigates.[1] These ships, based on a modified German MEKO design and built largely in Australia, began entering service in the mid-1990s and are scheduled to begin inactivation in the mid-2020s after 30 or more years of operational life. The SEA 5000 program was recently established to begin the process of inaugurating and managing the acquisition of the replacement for the *Anzac* class, termed the Future Frigate. As with any large, complex military acquisition program, SEA 5000 program managers need to understand the various paths and steps that could be taken in designing, building, testing, and supporting a complex naval ship and the implications of the various decisions a program faces along those different paths.

RAND is providing assistance to the SEA 5000 program by addressing the following research questions:

- What ship design and build options are available for the Future Frigate and what are the implications of choosing among the various options?
- What are the various phases, options, and decisions in a naval ship acquisition program?
- What are the important aspects that can contribute to the success of a program?

[1] The Royal New Zealand Navy also operates two *Anzac* frigates.

This report answers those research questions. Although specifically addressing the Australian SEA 5000 program, it provides guidance and support to any major ship acquisition program.

Design and Build Options

The SEA 5000 program has several options (materiel solutions) in terms of how it might acquire a new frigate capability to replace the *Anzac* class currently in operation. At one end of the spectrum, there is a *pure military off-the-shelf (MOTS)* option that will procure an existing foreign design, as is, for the RAN. The pure MOTS option could be built overseas or in Australia. At the other end of the spectrum is a *new design* (also known as a "clean sheet" design), which would be a class of ships specifically designed for the SEA 5000 program. It would allow for the most flexibility and customization by the RAN, but has the most risk and would require the most domestic resources to execute. Between these two options is a third option designated as *evolved MOTS,* whereby the RAN would adapt an existing design (working with the original design owner) and build the class mostly or entirely in Australia. This intermediate approach has a wide range of potential implementations that vary by the degree of change from the original baseline design. The more the baseline design is modified, the more risk there is.

Each acquisition option will have different influences on the flow of a program and the decisions required as the program progresses. In addition to laying out the roadmap for a major acquisition program, this report also describes how the different acquisition options affect the various points and decisions along that roadmap.

Ship Acquisition Process

This report provides a naval shipbuilding program overview that identifies the steps, decision points, milestones, and options along the various paths that can constitute a major acquisition program. The report examines the life cycle of a program from early development through operations and support. It highlights key areas that are different for naval ship acquisitions and important decisions that must be made along the way by illustrating with some recent examples of other ship or acquisition programs. It assumes that the reader has a basic knowledge of defense acquisition practice (e.g., understands such concepts as milestones, contract types, and cost and schedule estimates) but does not fully understand how the process needs to be tailored for naval ship acquisition.

Factors That Contribute to Program Success

All programs strive to be viewed as successful in delivering the desired operational capabilities on time and within budget. However, cost and schedule growth has been a significant issue for shipbuilding programs across the globe. Typically, first-of-class ships and submarines have been delivered later and at higher costs than was originally estimated. Although the increased complexity of modern ships has likely been a factor, cost and schedule growth have also been affected by management and execution issues throughout every phase of a shipbuilding program. As Australia embarks on a number of ambitious new shipbuilding programs, its shipbuilding budget and industrial base will face increasing challenges. Thus, it is more important than ever that program managers have the right tools at hand to manage program complexities and that the platform and capabilities are delivered within budget and on time.

Unfortunately, there is no magic solution to guarantee a successful program. Each new program will be different in some way and the options, decisions, and outcomes that surround a program will change. Even such countries as the United States and the United Kingdom, with long histories of designing and building new classes of ships and submarines and with relatively large defense budgets, have experienced difficulties in meeting cost and schedule goals for new programs. All programs experience some bumps along the way. Successful programs anticipate them and plan for their resolution.

This report describes the various factors that can influence a new program and how they have been addressed in previous programs. It draws heavily on the substantial body of RAND research on shipbuilding lessons learned and from root cause analyses of major defense acquisition programs.[2] The key lessons and factors are based on studies of numerous U.S., UK, and Australian ship and submarine programs. The report is intended to support the Commonwealth as it begins to plan for the SEA 5000 Future Frigate program.[3]

When considering lessons from the various programs, it is important to remember that the programs were conducted in different threat and budget environments and with evolving industrial bases for designing and building the military assets. Decisions were made on the basis of the environment at the time, so decisions varied by country and by program. Some lessons are unique to specific programs; others are unique to specific countries; some are universal. It is also difficult to judge the success or failure of program decisions. Views change during the conduct of a program and are based on the perspective of individuals. The important point is that the decisions were not necessarily "good" or "bad." Rather, they were or were not fully informed by knowledge of the risks and consequences.

Organization of the Report

Chapter Two defines the key attributes in establishing and managing a successful program. Chapter Three describes the various design and build options available for the Future Frigate and the key decisions that need to be made when evaluating those options. Chapter Four defines the various phases in a ship acquisition program and the steps, options, and decisions along those various phases. Chapter Five describes differ-

---

[2] The Bibliography lists RAND reports and other documents that provided insights into the management of previous programs.

[3] The discussion and recommendations in this report are geared toward management issues in shipbuilding and are not meant to address specific technical aspects of design or construction.

ent acquisition and contracting models and the advantages, disadvantages, and risks with each model. Chapters Six through Ten provide details on the various phases in a program life cycle defined in Chapter Four: Chapter Six describes the solutions phase, where requirements are set; Chapter Seven defines various options during design activities; Chapter Eight overviews the various steps and decisions in the manufacturing and build phase; Chapter Nine describes test and trials; and Chapter Ten describes the important issues with the in-service support of the Future Frigates. These chapters also provide key lessons and factors that can contribute to the success of a program. Chapter Eleven summarizes the key aspects of the previous chapters. Several appendixes provide additional details and information.

CHAPTER TWO

Establishing and Supporting a Program

Any new defense program has numerous stakeholders. They range from the Defence Materiel Organisation (DMO), which will acquire the new system, to the military service that will operate the system, to the minister, secretaries, and other political groups that budget for and support, or critique, the management and outcomes of the program, to the general public whose tax dollars fund the new acquisition. Many of these stakeholders have oversight and decisionmaking responsibilities.

The key organization for a new acquisition program is the program (or project) office. That office has a range of responsibilities and is faced with numerous decisions. It must assist in deciding what is needed and interact with the providers of the needed system on both a contractual and oversight basis. The program office must also interact with other stakeholders providing data, information, and recommendations on future paths. The program office typically starts small during the materiel solutions phase of the acquisition process but grows in size and capabilities as the program progresses. Many important contributions to program success are related to the development and sustainment of the program office.

But there are many stakeholders, each of which can contribute to program success or failure. Important lessons from previous programs center on the bigger enterprise, which includes the government, military services, politicians, and private-sector companies. Overall, the government that acquires the new system and the military service that operates and supports it must be knowledgeable of technologies, industrial base capabilities, and numerous other aspects of the decisions they will face during a major program acquisition. Basically, the government should have the capability to act as an informed and intelligent participant in the acquisition process.

This chapter describes the key factors surrounding the establishment and support of a program office that contribute to successful programs. Establishing and supporting the program office is key to a specific program, but the government must also realize that a single program is but one piece of an overall defense portfolio. Decisions made by other programs will have implications for the Future Frigate. Likewise, decisions made by the Future Frigate will affect other programs. On top of everything are the strategic considerations on such national issues as support for and development of an indigenous industrial base for naval warship design and construction.

Be an Intelligent and Informed Partner in the Shipbuilding Enterprise

An important lesson from previous programs is that the government, as with consumers of all products, must be intelligent and informed in its dealings with the private-sector organizations for the design, build, and support of a new naval ship. For the best outcomes, government must be closely engaged with the shipbuilder throughout the entire process. Therefore, to strengthen this message, we avoid the term "customer" here, since governments now understand that they must partner with the providers. Even though the government is the purchaser, the term "partner" better reflects the nature of the frequent interactions featuring a high level of information exchange that we recommend.

For the government to be an intelligent partner, its organizations need experienced technical personnel. Both the civilian and military sides of government should have centers of knowledge and expertise in such areas as hull dynamics, propulsion systems, signatures, combat and communications systems, and safety of operations. Many, if not all, of these knowledge centers should be in the government; however, academia and the private sector can augment or substitute in some technical areas.

To be an informed partner, the government must also understand past and current costs for the design and build of naval ships and be able to adequately estimate the cost of future ship design and build options. It needs to understand what factors drive costs and how different technical or managerial decisions can affect them. Collecting and organizing historical cost data, using the data to project future costs, and developing internal cost-estimating capabilities are needed to contribute to the cost-effectiveness analyses of various acquisition options.

There are several lessons from past and current programs where the government or military lacked the knowledge or foresight to make informed decisions. The UK's *Astute* submarine program began during a period in which the government, for budgetary reasons, dramatically reduced the technical resources that provided knowledge, expertise, and oversight for new ship design and build programs. Responsibilities for these technical capabilities shifted to the private sector, which was ill-prepared to accept them. As a result, the government was blind to the problems being faced by the prime contractor in developing the design and building the submarines.

The *Collins* submarine program is another example of how the government and military service did not adequately understand the implications of various decisions. The *Collins* was the first Australian submarine where the RAN had to assume the role of a parent navy.[1] For the first time, the government had to assume responsibility both for the build of the submarines and for their logistics support once they entered service. Support to the *Collins* fleet has fallen short of expectations during the first half of the life of the class.

A third example is the *Hobart*-class Air Warfare Destroyer (AWD), part of whose cost and schedule problems can be attributed to the government's overly optimistic view of the capabilities of the Australian shipbuilding industrial base.

The need for knowledge and expertise on the buyer's side does not really change for the different acquisition options. For a pure MOTS

---

[1] A *parent navy* operates and supports a ship or submarine that was largely designed and built in-country. Typically, that country is the only country that has that ship or submarine in its force structure.

option, technical and operational expertise is needed to evaluate the capabilities of available ship designs and determine how well they meet the desired operational requirements of the Future Frigate. Cost-estimating and schedule development expertise is also needed to evaluate trade-offs between what is available and what is needed and to predict future life-cycle costs for different build scenarios. These same skills and capabilities are needed to understand the cost, schedule, risk, and operational effectiveness implications of different types of evolutions to an existing design. Those skills also are needed, although to a greater degree, for a new ship design and build.

Delineate Roles, Responsibilities, and Decisionmaking Authority

The various roles and responsibilities in a new program basically come down to who should assume risks that arise. The assumption of risks by a particular entity should carry with it decisionmaking authority for that entity. The specific roles and responsibilities of the government and private sector and the locus of the final decision authority in various areas must be firmly established at the start of a new program. The responsibility for different risks should remain constant from program to program so that all organizations clearly understand how a new program will be conducted. However, circumstances may suggest moving responsibility for certain risks from the government to the private sector. Any such changes should be informed by a thorough analysis of how they might alter responsibilities and by a clear plan for the transition. Changes need to be adequately funded and the entities performing the activities need to be fully qualified to implement the changes.

Australia had to address this issue with the *Collins* and AWD programs. With the *Collins*, it proved difficult to determine which party had responsibility for certain risks and where final decisionmaking authority on design and build issues should reside. Programs in other countries also have experienced problems as a result of the assignment, or mis-assignment, of roles and responsibilities. For example, the UK experienced a major change in the locus of responsibility for certain risks at the beginning of the *Astute* program. Believing that the private sector could accomplish certain tasks at lower cost, many responsibilities previously held by government were transferred to the private sector. The role of design authority,[2] which had been filled by the Ministry of Defence (MOD) in previous programs, was assigned to the prime contractor. The private sector was ill-prepared to assume this new role and the two sides did not develop a plan for the transfer of responsibilities. The MOD adopted an "eyes on, hands off" policy (although with the drawdown of oversight resources at the shipyard, it effectively lost an "eyes on" capability).[3] The design authority role eventually reverted back to the MOD.

In some cases, the government must assume risks; in others, the prime contractor should assume them; finally, in many cases, risks should be shared. However, certain risks remain the sole responsibility of the government. These include obtaining the desired military performance from the new ship and ensuring safety of operations. Moreover, the acceptance of the ship as safe for operations is the responsibility of the government, and it should retain the role of design authority and assume the risks associated with the design authority process.

The government also should strive to deliver the overall program on time and within budget. It shares this risk with the prime contractor and must use all available tools to monitor contract performance, interact with the contractor, and optimally incentivize the builder to meet

---

[2] There are various "authorities" in a new program. For example, the U.S. Navy distinguishes between design authority and technical authority. The *design authority's* role is to forward to the designer the design specifications or rules. These are usually based on the ship concept selected from concept studies, which preceded the design effort. The design authority must be consulted and approve any proposed changes to the design specifications. In contrast, the *technical authority* is the subject-matter expert in various areas, such as the ship hull, mechanical and electrical engineering, ship safety, and ship design and engineering. The technical authority is responsible for establishing technical standards in each area and evaluating the risk if a design does not conform to technical standards during design and construction. To be effective, the design and technical authority roles require skilled and experienced staff with predominantly technical and engineering expertise.

[3] The fixed-price contract for the *Astute* program also tied the hands of the MOD. The MOD was reluctant to impose conditions or mandate changes to the design of the submarine for fear it would lead to cost increases.

schedule and cost milestones.[4] Part of the government's responsibility is also to be proactive in managing risks. It must identify where risks exist and develop a plan to mitigate them. And it must manage risks throughout the program—from initially setting requirements, through designing and building the ship, to accepting the finished vessel.

Because it will need to shoulder certain risks, the government should assume the following responsibilities:

- Set operational requirements for the new ship by working with industry, the navy, and other stakeholders.
- Assess safety and technical issues in accordance with the government's policy on safety risks.
- Oversee and monitor the design process to ensure that requirements and standards are met and, when necessary, provide concessions to those requirements.
- Oversee and monitor the build process to ensure that the ships are delivered on schedule and within projected costs.
- Ensure ship construction quality and acceptability by developing a testing, commissioning, and acceptance process so that the ships are delivered in accordance with the contract specifications and requirements.
- Ensure through-life safety and maintenance and postdelivery control of the design and construction of the ships in the class.
- Ensure that the model for logistics support fits the country's current and projected infrastructure for maintaining its ships and submarines.
- Ensure that program cost goals are reasonable and controlled.

Overall, the government and the private sector must establish an interactive partnership in which information flows freely and issues are discussed. Effective interactions will help the government better under-

[4] The prime contractor also faces risks if it does not efficiently deliver a cost-effective ship; however, although the prime contractor may go out of business, the government is still responsible for defense of the nation. Also, there are risks to the prime contractor if the ship is unsafe, but the government is ultimately responsible for the health and well-being of the sailors.

stand the product it will receive and help the prime contractor develop a product that better fits the navy's needs. An example of the need for effective relationships between the buyer and the seller was noted in the UK's National Audit Office (NAO) 2009 review of the Type 45 destroyer program, which suggested that poor working relationships between the government and industry partners led to program problems.[5] Joint governance arrangements between the MOD and industry did not have the appropriate mechanisms in place to resolve program issues in a timely way. During the program restructure, the government and industry developed new governance and communication arrangements that helped to better define roles and responsibilities. These arrangements also helped to establish processes for regular information sharing, progress reviews, coordinated planning, and quality reporting.

In the Future Frigate program, the RAN's and government's roles and responsibilities will vary for different acquisition options and acquisition phases. For a pure MOTS design, the roles related to ship design are held by the designing firm, but the government still has responsibilities for ensuring that the design meets safety standards and manufacturing quality (if built in-country). The government must also set requirements, establish testing procedures, and oversee construction activities. As a design evolves, the technical roles have to be shared by both the design firm and the government. With a new design and build option, the government assumes the majority of the technical roles.

Establish and Support a Program Office

The program office is the core of the government's role in an acquisition program. It has to assume or play a major part in many of the roles listed previously. It should be staffed with people of various skills and areas of expertise including ship technologies, safety standards,

---

[5] NAO, *Providing Anti-Air Warfare Capability: The Type 45 Destroyer,* London: The Stationery Office, 2009, p. 8.

construction oversight, logistics requirements and capabilities, cost and schedule estimation, and a host of others. The office starts with several people during the concept formulation stage and grows in both numbers and skills as the program progresses. Program office workforces for large programs range in size from several dozen to a few hundred depending on the roles of the program office and the extent of available personnel resources. It is not only the size of the core program office that is important, but the support provided by other public- and private-sector organizations. This is the point of the next lesson.

Involve All Appropriate Organizations

Although the program office should have core personnel with various knowledge and expertise, it also should have close technical and operational knowledge and ties to other public-sector or military service organizations. The program office will need to draw on those centers of knowledge as the program progresses. For example, it should have several military personnel from the RAN who are knowledgeable about operational requirements and tactics, logistics support, and crew training. Those RAN personnel should have the ability to reach back to their parent organizations for data, information, or analyses. Likewise, the program office should have a few dedicated cost analysts to gather data and build the models needed to inform various decisions. These cost analysts should have the ability to draw on resources from a central government cost-estimating group. Finally, to identify risks and solutions early and throughout the program, the program office must be supported by both research and technical personnel knowledgeable in a range of areas (hull, mechanical, and electrical systems and propulsion, signature, and survivability issues) and the construction shipyard(s), whose managers understand the potential problems that can arise from building certain aspects of a design.

One criticism of the *Collins* program was the absence of the technical community early on. Similarly, the UK's *Astute* program did not involve operators, builders, or maintainers to an appropriate degree during the early stages of the program. Some of the problems with these programs might have been alleviated if they had used a design and build philosophy—involving operators, maintainers, builders, and key suppliers—during the detailed design stages. Early involvement of builders, as well as operators and maintainers, not only helps identify requirements up front but also flags potential problems and their possible solutions.

The appropriate people and organizations in the Australian Parliament should also be informed of programmatic decisions and the status of a program. This is the focus of our next lesson.

Obtain and Sustain Political Support

A new defense program needs a range of supporters both inside government and the naval community and outside the program. Political support is necessary for the advancement and continued support of any new acquisition program. In the early concept development stage, RAN's needs must be clearly communicated; those needs should justify the operational requirements in terms of national priorities so that the government makes sufficient funding available to initiate and sustain the program. Because funding priorities may shift over time, the program office must also be aware of how its program fits into the larger defense portfolio, particularly how a particular ship's capabilities integrate or overlap with other platform capabilities.

Full disclosure during the program is necessary to obtain government, industry, and public support. There should be periodic feedback to government decisionmakers and to the public on how the program is progressing. Such feedback is especially important when there are unanticipated problems. In this regard, a good media management program is necessary. Effective communications with the press, academia, and government must be proactive, not reactive. One lesson from the *Collins* program is the need to effectively manage the media; the bad press that accompanied the *Collins* effort still taints the program in the mind of the general public. Program managers must proactively ensure that all parties are well informed in advance of positive and negative developments and their associated implications.

Develop Knowledgeable and Experienced Managers

Successful programs have experienced and knowledgeable people in key management, oversight, and technical positions. Growing future program managers and technical personnel within the civilian and military branches of government requires planning and implementation far in advance of any one specific program. Promising officers, especially engineering duty officers, and civilian personnel must be identified early in their careers and given suitable education and assignments to ongoing programs at a junior management level. Assigning people who have "earned their stripes" on one program is critical to the success of the next program.

Today, all countries, including the United States, face gaps during the start of new programs. Skilled personnel can face surges or contractions in demand for their skills, which makes it difficult to ensure that workforces with appropriate technical and managerial skills and capabilities are available when needed. However, all programs go through design, construction, and in-service support phases, so skilled personnel who face gaps in demand in one phase might be able to be reassigned to another phase to maintain their knowledge and skills. Key personnel could also be seconded to private industry or to defense laboratories when there are lulls in the demand for their specific skills. Another alternative is to define work that is not necessarily needed in the short term but could keep key personnel active while hopefully providing long-term benefits. This option, often viewed as "making work," should be weighed against the alternative of letting skilled personnel go and recreating the capability when needed.

Another important aspect is continuity in leadership and in team composition. Managers, leaders, and team members in the government and the industrial base should stay in a program long enough to gain knowledge of the program and maintain its goals. Frequent changes in leadership, which occurred in both the *Astute*'s prime contractor and the *Collins'* prime contractor and government program office, can degrade a program by introducing managers with different goals and strategies from those of their predecessors. Although personnel changes are inevitable, especially for military personnel, they should be minimized to the extent possible, and when new government or private-sector leaders are brought in, they should possess knowledge and experience similar to that of the individuals they replace.

Providing early experiences for future program managers is a challenge for Australia inasmuch as there are few opportunities for civilians and military personnel to gain expertise. Also, with low-volume, long-drumbeat[6] acquisition programs, the availability of experienced program managers may be limited. Many of the government personnel involved with earlier programs may retire and military personnel may be reassigned. Because of the shortfall in experienced personnel, Australia may require assistance from the ship design and build organizations of allied countries for future new programs.

Develop Realistic Cost and Schedule Estimates

Estimating costs and schedules in the early stages of a shipbuilding program are difficult tasks. However, the performance of a program is typically measured against early cost and schedule targets. Errors in cost-estimating have been found to account for approximately 15 percent of total cost growth that can be explained for major acquisition programs, whereas errors in schedule estimating have accounted for up to 10 percent of total cost growth.[7] Therefore, it is in the program manager's best interest to ensure that cost and schedule planning is conducted in a rigorous and systematic manner and that the cost and schedule targets that emerge are objective, realistic, and based on the best available information.

Developing robust cost estimates is not always straightforward. Lessons from other shipbuilding programs suggest that robust estimates require informed assumptions about technology risk, workforce composition and proficiency, and industrial base capabilities and should take into account the life-cycle costs of the program. Cost-estimating errors can result from using incorrect cost data or risk models. For example, estimating costs by benchmarking against legacy

---

[6] A production *drumbeat* is the duration between new ship construction starts.

[7] Joseph G. Bolten, Robert S. Leonard, Mark V. Arena, Obaid Younossi, and Jerry M. Sollinger, *Sources of Weapon System Cost Growth: Analysis of 35 Major Defense Programs,* Santa Monica, Calif.: RAND Corporation, MG-670-AF, 2008, p. 27.

systems may not capture the complexity of new technology and may underestimate costs. In addition, risk models should account for technical maturity and integration risks. In the DDG-1000 program, cost estimates at the time of Milestone B[8] did not fully take into account the uncertainty surrounding the level of effort needed to mature and to integrate critical technologies, leading to cost and schedule breaches.[9]

One aspect of setting realistic cost estimates is having the best possible knowledge available at the time of the estimate. Current policy for major U.S. acquisition programs requires that the confidence level for baseline cost estimates be at least 80 percent.[10] In shipbuilding, best practice suggests that construction of the lead ship should be priced only when the detailed design is sufficiently complete to provide enough knowledge for realistic cost estimates.[11]

Some initial cost and schedule estimates for major acquisition programs are often found to be overly optimistic on later review.[12] Reasons for this may vary. Target costs may be deliberately underestimated so that a program looks more attractive to the legislative body that approves the funding or so that it is not viewed as "too expensive" to

---

[8] Milestone B is the point in a U.S. program considered the official program start.

[9] Irv Blickstein, Michael Boito, Jeffrey A. Drezner, James Dryden, Kenneth Horn, James G. Kallimani, Martin C. Libicki, Megan McKernan, Roger C. Molander, Charles Nemfakos, Chad J. R. Ohlandt, Caroline R. Milne, Rena Rudavsky, Jerry M. Sollinger, Katherine Watkins Webb, and Carolyn Wong, *Root Cause Analyses of Nunn-McCurdy Breaches, Vol. 1: Zumwalt-Class Destroyer, Joint Strike Fighter, Longbow Apache, and Wideband Global Satellite,* Santa Monica, Calif.: RAND Corporation, MG-1171/1-OSD, 2011, p. 26.

[10] U.S. House of Representatives, 111th Congress, 1st Session, *Weapons Systems Reform Act of 2009,* Public Law 111-23, Washington, D.C.: Government Printing Office, 2009.

[11] John F. Schank, Cesse Cameron Ip, Kristy N. Kamarck, Robert E. Murphy, Mark V. Arena, Frank W. Lacroix, and Gordon T. Lee, *Learning from Experience, Volume IV: Lessons from Australia's Collins Submarine Program,* Santa Monica, Calif.: RAND Corporation, MG-1128/4-NAVY, 2011a.

[12] Irv Blickstein, Jeffrey A. Drezner, Martin C. Libicki, Brian McInnis, Megan McKernan, Charles Nemfakos, Jerry M. Sollinger, and Carolyn Wong, *Root Cause Analyses of Nunn-McCurdy Breaches, Volume 2: Excalibur Artillery Projectile and the Navy Enterprise Resource Planning Program, with an Approach to Analyzing Complexity and Risk,* Santa Monica, Calif.: RAND Corporation, MG-1171/2-OSD, 2012, p. 82.

move forward as currently specified.[13] Although these initial estimates may secure funding and support for the program, they will most likely cause issues later if the program experiences cost overruns. Overly ambitious schedule targets may reflect a need to replace aging vessels or to maintain the shipbuilding industrial base. Again, if the program is unable to deliver on unrealistic schedule targets, it may suffer reduced political support. In addition, cutting corners to meet schedule targets may compromise the quality and capability of the ship's systems. It is important to recognize "optimism bias" when developing cost and schedule estimates and ensure that there are procedures for mitigating the effects of any types of bias.

Unrealistic cost estimates may also also result from the inexperience of the cost-estimating staff, inadvertent errors in cost estimation, or flawed assumptions. For example, the feasibility of meeting performance expectations is often overestimated in shipbuilding programs. In the DDG-1000 program, planned capabilities relied on technologies that had not been demonstrated in large surface combatants. There was limited historical evidence to support the feasibility of achieving these performance targets and thus little or no benchmark data to support robust cost and schedule estimates.[14]

Another area where shipbuilding programs have been overly aggressive in performance targets is in reduced crew complements. A smaller complement can result in reduced costs through fewer berths, less storage space, and reduced hotel service capacity (for example, fresh water production capacity). However, unrealistic reductions in crew numbers can result in shortfalls in operational effectiveness or material readiness. The U.S. Littoral Combat Ship (LCS) program is an example of a program where the initial crew complement was underestimated. During trial operations, it was found that 20 crew members would need to be added to the planned 75-member crew to effectively conduct counternarcotics operations.[15] Although it is relatively inex-

---

[13] Bolten et al., 2008, p. 17.

[14] Blickstein et al., 2011, p. 24.

[15] Philip Ewing, "20 to Join LCS Crew on Trial Deployment," *Navy Times,* November 14, 2009; and U.S. Government Accountability Office (GAO), *Littoral Combat Ship: Actions*

pensive to add berthing space in the design phase, the LCS program is now incurring higher costs through rework of existing platforms. Cost analysis suggests that planning for a larger crew may have a relatively small effect on total cost if accounted for early in the program and may offer more flexibility for future mission capability while avoiding the need to retrofit the platform.

Another part of developing realistic estimates is making robust assumptions about shipbuilder workforce learning and productivity. A rapid buildup of the workforce for a new shipbuilding program may affect the quality-assurance process, particularly if the workforce does not have the necessary skills and experience. For example, in the *Ohio*-class submarine program, a number of welds were faulty and required costly rework. The faulty welds were found to be correlated with a rapid buildup of the workforce and inexperienced welders and quality-assurance managers.[16] In Australia's AWD program, the risks of re-establishing Australia's capacity to build warships were underestimated and expected efficiency gains built into the initial cost models were not realized.[17] The program office should account for the workforce buildup and apply realistic learning curves when developing estimates.

Understand the Effects of External Factors Beyond the Program's Control

Program managers have some degree of internal control over cost and schedule during the design and build phases, but some external forces are not under their control and can affect program success. Three of these forces that have had an effect on program cost or schedule in past

---

Needed to Improve Operating Cost Estimates and Mitigate Risks in Implementing New Concepts, Washington D.C.: U.S. Government Printing Office, February 2010.

16 John F. Schank, Cesse Cameron Ip, Frank W. Lacroix, Robert E. Murphy, Mark V. Arena, Kristy N. Kamarck, and Gordon T. Lee, *Learning from Experience, Volume II: Lessons from the Ohio, Seawolf, and Virginia Submarine Programs,* Santa Monica, Calif.: RAND Corporation, MG-1128/2-NAVY, 2011b.

17 Australian National Audit Office, *Air Warfare Destroyer Program; Audit Report No. 22 2013–14,* Canberra, Australia, 2014, p. 51.

shipbuilding programs are government funding decisions that affect quantity or schedule, changes to the health of the industrial base, and economic matters related to exchange rates or inflation. Program managers need to understand these external factors to recognize warning signs for program cost or schedule growth and to take steps to mitigate risks.

Government Funding Decisions

Funding stability is important for the effective management of a program and is a factor to consider regardless of the acquisition strategy pursued by the SEA 5000 program. Some government decisions beyond the control of the program manager may cause program instability and increase the risk for cost growth. These decisions may be due to shifts in national priorities or in the fiscal environment. For example, these may be decisions relating to

- quantity—increase or decrease in the number of platforms needed
- schedule—increase or decrease in the schedule length for production.

RAND has found that up to 22 percent of total cost growth for major defense acquisition programs can be explained by government quantity changes, and 9 percent of growth can be explained by government schedule changes.[18] Quantity changes in shipbuilding programs may arise from changes in fleet size requirements, in capability needs, in defense priorities, or in the funding environment. For shipbuilding programs, it has historically been more likely for the government to reduce the production quantity than to raise it. When ship quantities are reduced, the program's initial assumptions about various elements of life-cycle costs need to change. For example, the per-unit research, development, test, and evaluation (RDT&E) costs are typically higher if they are spread over fewer platforms, whereas other program costs related to operations and maintenance (O&M) or life-cycle maintenance may increase or decrease depending on the efficiencies lost by

[18] Bolten et al., 2008, p. 27.

reducing the number of hulls in production. In addition, quantity changes may affect the prime contractor's ability to leverage long-term relationships with its suppliers, particularly for components that are unique to the platform.

At the outset, the DDG-1000 program was expected to deliver 32 ships at an approximate program acquisition unit cost (PAUC) of $1 billion.[19] As the desired sophistication (and cost) of the vessel increased, the number of ships that the Navy could produce under the existing funding environment fell, dropping to ten by 2005 and to three by 2009. The RDT&E baseline costs in 2005 accounted for approximately 26 percent of total program costs; after the reductions in quantity, RDT&E accounted for just over 50 percent of total program costs. As of 2013, the U.S. Navy plans to procure only three vessels at an average procurement unit cost (APUC) of over $3 billion and a PAUC of $6.3 billion per vessel.

Decisions from the government to compress a program schedule may be a result of the need to field a capability earlier. Alternatively, government decisions to extend a program schedule may be in response to fiscal pressures requiring that funding be spread out over a greater period of time. For example, part of the motivation for extending the AWD delivery schedule in 2012 by approximately two years was to reduce demand on the Commonwealth budget.[20] Schedule changes by the government will most likely have cost and potentially contractual implications. Unexpected schedule changes can also affect shipyard workloads and the ability of contractors to develop or maintain an appropriate quantity of skilled workers. However, in the case of the AWD, the extended schedule was intended to support a more stable workload and to avoid peaks and troughs in the shipyard workforce.[21]

---

[19] The PAUC is calculated as the total of (RDT&E + procurement + unique military construction) ÷ (total procurement quantity + RDT&E prototypes). The APUC is calculated as the total procurement dollars ÷ total procurement quantity.

[20] Australian National Audit Office, 2014, paragraph 1.23.

[21] Australian National Audit Office, 2014, paragraph 1.23.

Industrial Base Issues and Market Conditions

The health of the industrial base, including the financial strength of the prime contractor, shipyards, vendors, and other subcontractors, can all affect program cost and schedules. Assumptions made in the initial phases of a program about the health of the industrial base can result in overestimating labor productivity and underestimating capital investment needs. In addition, assumptions about what products will be available commercially or competitively can also result in an over- or underestimation of labor productivity and capital investment needs.

During the design phase for the AWD program, for example, the government derived cost and schedule estimates on the assumption that subcontractors had the financial capacity, facilities, and commercial incentive to develop the capabilities needed to execute contracts for the hull block production. During tendering, however, it became apparent that the shipyards would require significant capital investment to develop the necessary capabilities.[22] Other major acquisition programs in the United States also have made optimistic assumptions about development in the commercial sector. For example, they have overestimated the readiness of three-dimensional design and manufacturing software for ship and aircraft construction and the development of commercial satellite launch capabilities.

Shipyard management issues or labor disputes also can affect cost and schedule performance. During the construction of the *Seawolf*- and *Ohio*-class submarines in 1988, the Metal Trades Council union staged a 21-day strike, which contributed to a two-month delay in completion of the *Ohio* lead ship.[23] In the *Virginia* program, increases in labor hours—which accounted for approximately 40 percent of cost growth—were a result of union strikes and inefficiencies in integration between the two construction shipyards.[24] Although program managers cannot control these issues, it is important that they remain aware of potential issues through an onsite shipyard presence and engage in frequent dialog and relationship building with shipyard counterparts.

---

[22] Australian National Audit Office, 2014, p. 26.

[23] Schank et al., 2011b, p. 25.

[24] Schank et al., 2011b, p. 87.

Economic Matters

Unforeseen economic matters may affect a program's ability to initially make robust estimates and can lead to unexpected cost increases across all phases of shipbuilding. This includes program cost changes associated with differences between predicted and actual inflation. If many of the ship's components are to be sourced from international suppliers, exchange rates can also be a big factor in rising costs. For example, the Australian National Audit Office found that for the AWD program, the combined effects of inflation and foreign exchange variations between June 2007 and December 2013 resulted in budget increases of $722 million.[25] Although fluctuations in inflation and exchange rates are not under program manager control, the best predictions should be incorporated into the cost models, and cost estimates should be routinely updated to reflect change.

Take a Long-Term, Strategic View

The SEA 5000 program will need to take into account that the Future Frigate is but one piece in the overall RAN portfolio. Decisions for the Future Frigate will affect and be affected by other RAN programs.[26] This is especially true for the development, use, and sustainment of the ship and submarine industrial base that designs, builds, and supports the RAN fleet. The Commonwealth is grappling with the level of capability desired from the industrial base and how best to sustain that

---

[25] Although inflation raised the cost estimate by $1.183 billion, appreciation of the Australian dollar decreased the estimate by $451 million, which created a net effect of $722 million. See Australian National Audit Office, *2012–13 Major Projects Report,* Canberra, Australia, 2013b.

[26] A new ship acquisition program may depend on other programs to provide various systems that will be installed on the ships. There are advantages of centrally managing certain combat and weapon systems, including economies of scale in acquisition and support. For example, in the United States, there are Program Executive Officers for Integrated Warfare Systems and for command, control, communications, computers, and intelligence (C4I) systems. These centralized organizations manage the procurement and in-service support of systems that cross multiple ship classes. For major system management, they work best when there is sufficient demand across various ship classes.

capability. Once clear guidance is provided, the Future Frigate program can make informed decisions on the acquisition path to follow—pure MOTS, evolved MOTS, or a new design. The program must also interact with other new design and build programs to determine the most cost-effective way to acquire and support new ships and submarines that meet the objectives of national policy.

There is also the need to sustain government technical expertise after the ships enter service. Maintenance procedures may change and issues may arise that require knowledge of system designs. Also, a new ship does not remain static once it is delivered. Technologies change, new capabilities are needed, and new threats emerge and evolve. These evolutions require experienced designers and engineers to maintain a technology and capability edge and to update existing platforms with new technologies and new capabilities.

Australia developed a submarine construction capability with the creation of the ASC.[27] But the country had no plans on how to sustain that capability once the *Collins* boats were built. In the UK, the substantial gap between design and build of the *Vanguard* class and the start of the *Astute* program was a big contributor to the problems faced by the *Astute* program. This led to a situation in which submarine design and build skills atrophied, resulting in a costlier and lengthier *Astute* procurement effort. The issue is not that the gap should have been avoided but that the MOD neither anticipated the effect of the gap nor factored the need to rebuild its industrial base capability into the cost and schedule estimates.

In the future, similar gaps are likely because of constrained defense budgets and the long operational lives of ships. Governments must decide at what level to sustain sufficient resources and expertise during those gaps to allow reconstitution when needed. There are costs and benefits of sustaining various levels of skilled and experienced resources.[28] In addressing these options, governments must be prepared

---

27 ASC was formerly known as Australian Submarine Corporation.

28 See John F. Schank, Jessie Riposo, John Birkler, and James Chiesa, *The United Kingdom's Nuclear Submarine Industrial Base, Volume 1: Sustaining Design and Production Resources*, Santa Monica, Calif.: RAND Corporation, MG-326/1-MOD, 2005b; John F. Schank,

to estimate the implications of a gap on future programs and the cost of sustaining resources during a gap.

Key Points in Establishing and Supporting a Program

- *Be an intelligent customer who understands the implications of various decisions and an informed customer who knows the status of programs.* Ensure that new processes and new systems are fully analyzed and not just theoretical ideas.
- *Delineate the roles and responsibilities of the DMO, private contractors, and subcontractors.* Design authority, or at least detailed knowledge of the design and authority to maintain and modernize systems, should reside with the government. If major responsibilities are shifted from the government to the private sector, it is important to ensure that industry is qualified to accept those new responsibilities.
- *Develop knowledgeable and experienced managerial, oversight, and technical support personnel.* Growing future program managers and technical personnel within the DMO and the RAN requires planning and implementation far in advance of any specific program.
- *Develop realistic cost and schedule estimates.* Assumptions driving cost and schedule estimates should be unbiased and based on the best available knowledge.
- *Take a long-term, strategic view of the naval fleet and the industrial base.* Understand how a specific program affects the long-term strategic plan for the fleet and the supporting industrial base.

Cynthia R. Cook, Robert Murphy, James Chiesa, Hans Pung, and John Birkler, *The United Kingdom's Nuclear Submarine Industrial Base, Volume 2: Ministry of Defense Roles and Required Technical Resources,* Santa Monica, Calif.: RAND Corporation, MG-326/2-MOD, 2005c; and John F. Schank, Mark V. Arena, Paul DeLuca, Jessie Riposo, Kimberley Curry Hall, Todd Weeks, and James Chiesa, *Sustaining Nuclear Submarine Design Capabilities,* Santa Monica, Calif.: RAND Corporation, MG-608-NAVY, 2007.

CHAPTER THREE

SEA 5000 Acquisition Options

A major program goes through three life-cycle phases: design, construction, and in-service support. The potential acquisition paths for the Future Frigate program involve various options across these three phases. Design options range from using an existing design with no (or very little) modification to developing a new ship design. A spectrum of options spreads between these two extremes depending on the degree of the desired change to an existing design or on how revolutionary a new design may be compared to existing ship designs. Modifications to an existing design could include major structural changes, increases in ship service margins (such as power, cooling, and bandwidth), or the use of different mission systems. As the size and the scope of the desired changes to an existing design grow larger, costs, schedules, and risks may approach, if not exceed, those associated with a new design. Important considerations when evaluating ship design options are the match of desired requirements to available capabilities and the availability and competence of design resources.

Construction options range from building the ships in another country to building them in Australia. As with design, some options, such as a shared build between countries or shipyards, fall between these two extremes. An important aspect of the ship construction phase is testing and evaluating the completed ship. The more a ship's design deviates from an existing design, the greater the cost and lengthier the schedule to conduct operational tests and remediate any deficiencies. The proficiency of the construction shipyards and the policies for the

Australian shipbuilding industrial base will factor into the evaluation of construction options.

Although acquisition decisions typically focus on the design and construction of a new class of ships, it is equally important to consider how a new ship will be supported during the operational life of the class. A new naval combatant will typically operate for 30 or more years. It will require *upkeep,* or the maintenance and repair of ship and weapon systems, so that it operates according to specified capabilities. Thirty or more years is a long time, and desired operational capabilities of the Future Frigate will change to meet new missions, adversaries, and technologies. Such changes, especially those that involve technology and manufacturing advances, will lead to the obsolescence of certain equipment or components and the need to *update* existing equipment to keep pace with changes in the supply of parts. Technical obsolescence involves both hardware and software. Given the high degree of computer-related functions on a complex naval ship, decisions to no longer support a specific software system can result in costly and time-consuming updates. Moreover, future technologies and new missions may lead to major system and equipment *upgrades*. System upgrades move beyond updates by replacing major equipment and systems with new models rather than just replacing old components with new ones on existing equipment and systems. The adaptability features of a design are important considerations when choosing a pure MOTS or evolved MOTS option.

The long operational life of naval combatants and the uncertainty of how the basic ship must change during that long life require careful considerations and planning for the upkeep, update, and upgrade of the Future Frigate. Thus, as early in the program as possible, it is important to develop the concept of operations (CONOPS) for the Future Frigate that defines how the ship will be supported and operated. Part of the CONOPS is an integrated logistics support (ILS) plan. This ILS plan must address:

- the development of a maintenance and modernization plan that identifies specific maintenance periods and spells out the objec-

tive, duration, and estimated workloads involved in each such period
- the technical data and permissions to allow the upkeep, update, and upgrade of the ship to be conducted in Australia by Australian organizations in accordance with the ILS plan
- relationships with the equipment suppliers to support their products or to provide the knowledge, expertise, and information on their products to another Australian organization
- the desired modularity and flexibility aspects of the Future Frigate design to provide the ability to adapt to future changes in a cost-effective manner.

A range of technical data and expertise is needed to maintain and modernize the ships over the life of the class. Effective in-service support requires some knowledge of the ship design, including access to equipment specifications, system capacities and limits, hull fabrication specifications, detailed naval architecture weight and balance data, and a host of other technical data. This information would be necessary to make informed decisions about whether specific upgrades to the Future Frigate would be possible or would be limited by its initial design. These and other types of data are typically contained in a three-dimensional computer product model. The availability of this product model and its ability to support maintenance and modernization are important considerations when evaluating acquisition options. The issues of technical data availability lessen with a new design ship.

When defining the various acquisition options for the Future Frigate, it is informative to start with the two extreme end points. At one end is using an existing design of another country's warship and having those ships built outside Australia. One could liken this acquisition path to buying an automobile built in another country. The design is fixed; various options to the base package are available; and a complete, tested car is delivered to the customer. This option basically involves deciding which existing foreign design best meets the desired operational requirements for the Future Frigate and contracting with the overseas manufacturer to deliver the completed ships to Australia. We call this the "pure" MOTS option. The acquisition of the

Oberon-class submarines from the UK is an example of this, wherein the United Kingdom built and supported the ships for the RAN and acted as the parent Navy.[1]

At the other end of the acquisition spectrum would be designing a new ship in Australia and constructing the entire class there. The *Armidale* patrol boat is an example of an Australian (relatively) new designed and built naval ship.

It is unlikely that either of these two extreme options will apply to the acquisition of the Future Frigates. Even if an existing design is used, some level of design modification is almost always required, and ship construction might involve shipyards inside and outside Australia. This has been the case in many Australian ship programs, including the *Collins*-class submarines, the *Anzac* frigates, the AWD, and the new Landing Helicopter Dock (LHD) project. Also, the Commonwealth has the strategic goal of developing the Australian shipbuilding industry as a national security asset.[2] The other extreme option of a new design also has low probability because of the time, cost, and risk of a new design, especially when there are insufficient levels of complex naval ship design resources within Australia. Although these two extremes may have very low probability of being viable candidates for the SEA 5000 program, describing them can serve as anchor points for descriptions of the more likely acquisition options.

Pure MOTS Option

At one end of the acquisition spectrum would be the option of the SEA 5000 program using an existing ship design with little or no modification and having those ships built in another country. This was the path

---

[1] For the *Oberon* class, the Royal Navy provided training and maintenance and modernization support for the Australian submarines and those of other countries. For the *Collins* class, the RAN had to logistically support a class of submarines with little or no help from another country. Thus, the RAN had to assume the parent role for the class.

[2] Australian Department of Defence, Defence Materiel Organisation, *Future Submarine Industry Skills Plan: A Plan for the Naval Shipbuilding Industry,* 2013. Also see DefenceSA, *Naval Shipbuilding: Australia's $250 Billion Nation Building Opportunity,* undated.

taken in the early stages of the Australian submarine program where *Oberon*-class submarines were provided by the UK. Currently, several frigate-sized ships from other countries are potentially available. There are also allies in the early stages of new, small surface-ship combatant programs, including the United States (follow-on to the LCS class), the UK (the Type 26 program), and Canada (replacement for the *Halifax* class). Typically, some options, such as different communications and weapon systems, are available with an existing design. Germany's MEKO-class ships were sold to several countries, including the *Anzac* program with slight variations in capabilities.

Rarely is an existing design adopted "as is." Some type of change is almost always needed, if not for different operational capability, then for differences such as environmental constraints or ship crew accommodation standards. So, even with a pure MOTS design, some design work will be required to tailor the ship to Australia's needs. The line between a pure MOTS option and an evolved MOTS option is not always clear. Here, we define a pure MOTS option as involving no major structural changes to the ship's hull, mechanical, and electrical (HM&E) systems or to the various weapon and combat systems offered by the host country. The existing design could include the ability to choose between different gun or missile systems, for example. This is similar to an automobile manufacturer offering different audio systems or luxury packages with a given model. But, to be a pure MOTS option, those different system options and their integration into the ship should have already been designed, tested, and, hopefully, built.

Any acquisition strategy entails advantages and disadvantages connected with costs, schedules, and risks. The pure MOTS option should provide a proven design and a continuous construction process that has become efficient when producing ships. Designs that are in the early stages and that have never been built and operated carry significant risks. First-of-class ships typically cost more and take longer to deliver than suggested by early program estimates. Costs, schedules, and risks connected with a proven design should be lower than other acquisition strategies. However, the capabilities of ship designs available via the MOTS path are likely to vary from the operational requirements that the RAN may desire for the Future Frigate. Trade-offs and compromises between cost and capability are often needed when choosing between available ship alternatives.

Another potential disadvantage of the pure MOTS option is that it would not sustain and grow Australia's ship design and construction industrial base. Although resources would be needed to manage the program and to gain a basic understanding of the ship design, the pure MOTS path would place little or no demands on ship construction resources. But the effects on the shipbuilding industrial base of a pure MOTS option for SEA 5000 should not be considered in isolation from other programs. It is important to understand the magnitude and timing of the demands on various design and construction resources from all programs when deciding how to best sustain Australian shipbuilding capabilities.

The pure MOTS option could change slightly to include building some or all of the ships in Australia. For example, the *Collins* program involved a shared build between Sweden and Australia, and the current LHD and AWD programs involve a shared build with Spain and several shipyards in Australia. This would help support ship construction but will not grow new ship design capabilities.

Another important consideration when evaluating the pure MOTS option is in-service support and modernization. Although some support options and warranties may be available, the Future Frigate will be based and supported in-country. Therefore, it is important that the pure MOTS option include the technical data needed to repair and maintain the ship and her systems and the technical approval to update equipment and systems as needed during the operational life of the class. Also important is the degree of adaptability of the ship design. Some ships—such as the U.S. Navy's LCS and the Royal Danish Navy's *Absalon* class—provide the flexibility to change mission modules relatively quickly.

New Design, Build in Australia Option

The other end of the acquisition spectrum would be to design and construct a new ship class wholly within Australia. In all probability, this acquisition path would involve the longest timeline and the greatest cost and risk, depending on how revolutionary a design is desired. The advantage of a new design is its ability to meet the RAN's desired operational capabilities for the ship, including survivability, availability, and adaptability. This path also does the most to grow and sustain Australian ship design, construction, and in-service support resources and could eliminate all issues and concerns surrounding technical data rights.

One hurdle for a new design Future Frigate is the availability of design resources in Australia. Australia has never designed a complex naval combatant on its own. New ship design programs require hundreds if not thousands of skilled personnel, computer design resources, and various facilities. Growing these design capabilities in Australia would take time and money. The more likely path for a new design effort might be a partnership that involves design houses from another country. These partnerships would likely include Australian and foreign shipbuilding and combat system companies.

A potential advantage of a new design could be the ability to sell the new ship to other countries through export agreements. Many emerging nations desire frigate-sized combatants for their naval forces. However, the market for export sales is crowded with numerous potential sellers.

Evolved MOTS Options

An evolved MOTS option would involve making some fairly significant changes to an existing ship design. Changes could include incorporating structural modifications to increase or reduce the ship's size and displacement; altering the ship's power, cooling, and bandwidth capabilities; or using different mission equipment and systems. The evolved ships could be built entirely or partially in the country that owns the ship design or entirely in Australia.

Numerous alternatives are available to the SEA 5000 program under the evolved MOTS acquisition path. Several U.S. and European shipbuilders are proposing variations of their existing ship designs to meet frigate-type requirements.[3] As mentioned, Canada is beginning deliberations on a replacement for its *Halifax*-class frigates, the United States is looking at a small surface combatant, and the UK's Type 26 program is under way. Depending on the timing of the allies' programs, opportunities for partnerships may exist. The AWD or the *Anzac* frigate basic hull form also could serve as a basis for the Future Frigate.

As with a pure MOTS option, the evolved MOTS option has advantages and disadvantages. If modifications to the basic design are not too substantial (if, for example, they involve no major structural changes), then design schedule and cost—although greater than for a pure MOTS option—will be less than for a new design, especially if an existing product model for the ship requires only minor updates. Also, a warm production line can lead to reduced construction time and costs compared to a new design option. Finally, technical risks are lower than with a new design.

As the magnitude and scope of the changes to the basic design grow, so too do costs, schedules, and risks. At some point, the costs, schedules, and risks grow to potentially meet or exceed those for a new design and build. Again, trade-offs will be needed between the cost and effectiveness of various evolved MOTS alternatives, especially the degree of desired change to the basic ship design.

To summarize, the various acquisition options available to the SEA 5000 program have different advantages and disadvantages that affect the operational capabilities of the Future Frigate as well as the costs, schedules, and risks to the program and the effect on the Australian shipbuilding industrial base. Careful and thorough analyses will be needed as the program progresses to examine the cost-effectiveness trade-offs among those various options.

---

[3] For example, Huntington Ingalls Industries in the United States is offering a variant of the *Legend*-class National Security Cutter that it is building for the U.S. Coast Guard, and Lockheed Martin is offering a modified *Freedom*-class LCS platform currently being built for the U.S. Navy.

Important Considerations with Different Acquisition Options

As described above, the SEA 5000 program can consider a wide spectrum of acquisition options. Deciding which option provides the most cost-effective path will require several analyses of the alternatives. These analyses—and the future policy deliberations that they inform—must consider, and will be influenced by, constraints and strategic policy objectives.

Operational Requirements

When considering a pure MOTS acquisition path, the program must determine if any existing and available designs can meet the desired operational requirements of the Future Frigate. If no available design meets desired operational requirements, the program then must determine how seriously those shortfalls might affect operations and how existing designs could be modified to eliminate them. The decision to adopt an evolved MOTS option will require trade-off analyses between the operational capabilities and cost, schedule, and risk. Even evolved MOTS options will have some limits on the types of equipment and systems installed on the ship—especially in the command, control, communications, computers, intelligence, surveillance, and reconnaissance (C4ISR) area.

Schedule Constraints

The desired date to introduce the Future Frigate into the operational fleet could constrain some acquisition options. It can take ten years or longer to design, build, and test a new surface combatant. The time between concept and the delivery of the first-of-class could be even longer if the new design involves significantly different features and capabilities than any existing ship design and if the industrial base is not resourced and prepared to efficiently produce the lead ship. A tight schedule could preclude a new design option or even an evolved MOTS option that requires major modifications to the base ship design.

Program Costs

Costs are typically a prime determinant when selecting an acquisition option. When evaluating acquisition options, it is important to understand the total life-cycle costs, including those incurred by the government and private sector for the ship design, construction, and in-service support. Understanding the trade-offs between costs and capabilities is also important. Such an understanding derives from an appreciation of the relationship between cost and such operational measures as speed, survivability, endurance, and mission effectiveness that will make up the ultimate operational capabilities of the Future Frigate. Early in a program, cost estimates are typically rough-order-of-magnitude (ROM) values considered at a fairly high work-breakdown level (e.g., single digit). These ROM cost estimates should be confidence intervals versus point values. They may lead to the elimination of certain acquisition options and narrow the field of choices. As the program progresses, ROM costs are refined with additional data and insights, and cost elements are estimated at a lower work-breakdown level.

There will some design-related costs even with a pure MOTS option. If all or parts of the ships are built in Australia, the existing design is used to create a detailed design specific to the construction shipyard(s). The detailed design takes into account the facilities, manufacturing equipment, and capabilities of the shipyard and produces "blueprints" of the build process for that yard. The detailed design, manufacturing specifications, supply lists, and other construction data are contained in an electronic product model created using three-dimensional computer-assisted design and manufacturing software packages. The availability of the product model and the software platform to operate it are important considerations when considering pure or evolved MOTS options.

Specific program costs should not be the only focus. The total costs of the portfolio of current and future Australian ship and submarine programs should factor into the acquisition decisions for the SEA 5000 program. Choosing an evolved MOTS or a new design option could affect the demand for and availability of design resources for other programs. Choosing to build the ships in whole or partly in other countries will affect the ship construction costs of other current and future programs.

Technical Risks

All programs seek to eliminate or reduce technical risks, inasmuch as higher risks typically lead to increased cost and schedule delays. A pure MOTS option should carry the lowest technical risks, given that it would involve a proven ship design. This is especially true if the pure MOTS option were to use existing ship equipment and combat systems. Technical risks with evolved MOTS options typically grow as the desired design modifications become more substantial. Usually, a new design will carry the greatest technical risks, especially if systems on the new design are not available or proven.

In addition to technical risks, industrial base risks are involved with the construction of the ship. New manufacturing techniques or new types of materials can lead to increased construction costs. Also, an industrial base that is not resourced and prepared to build complex warships can lead to production inefficiencies and increased costs and schedules. Most first-of-class ships take longer to build and cost more than originally estimated.

Desire for Competition

Competition is viewed as one way to reduce program costs. Competition can be applied at different levels—for ship design, for ship construction, or for the equipment and systems used on the ship. A pure MOTS option can incorporate competition among existing designs that can meet the desired operational requirements for a ship class. An evolved MOTS design may not involve competition for the ship design but could use competition for ship construction, major weapon systems, and vendor-supplied equipment. A new design effort could involve competition among potential partners that could help Australian companies design the ship and shipyards that could build all or parts of the ship.

The program must weigh the potential benefits and costs of any competitive awards. Conducting a competition can incur costs to provide funds to two or more companies. Also, competition may result in one or more companies leaving the naval ship construction industrial base, thus precluding future competitions.

Industrial Base Policies

The choice of acquisition path for the Future Frigate must be informed and shaped by the strategic goals for the Australian shipbuilding industrial base. An Australian policy of building and sustaining shipbuilding capabilities will challenge decisions to use a pure MOTS option that relies on constructing ships outside Australia. Again, all current and future Australian shipbuilding programs must be considered when making decisions that potentially affect in-country shipbuilding design and production resources.

In-Service Support and Modernization

How the ship will be supported—including the upkeep, update, and upgrade of equipment and systems—is an important consideration when selecting an acquisition option. A pure MOTS option will present challenges in supporting the ship class during its life unless adequate agreements are in place with the ship design organization and the companies supplying the major equipment and systems for technical support, data rights, and approvals for system modifications. The majority of a ship's total life-cycle cost comes after the ship is delivered. Personnel, maintenance, and modernization costs over the 30-plus years of service can far exceed the costs of designing and building the ships. Any existing, evolved, or new design should have modularity and flexibility features that permit the shape to adapt to future changes in technologies and missions.

Summary Comments

A range of acquisition paths is available to a new naval ship program. These paths vary from buying an existing ship on the open market to designing and building a whole new class of ships. There are numerous alternatives between these two broad options. The more likely scenario, based on recent Australian shipbuilding programs, is to modify an existing design to some degree and to build the ships at least in part in Australia. Pure MOTS ships often do not meet Australian requirements, and contracting arrangements throughout the life of the program can prove challenging. New designs require numerous ship design resources that take time and funds to develop. Constructing a class of eight frigates requires capacity and proficiency in the shipbuilding industrial base. Even with the hurdles associated with pure MOTS or new design ships, those options must be evaluated early in a new program. It is the early decisions that shape the program and factor into its success or failure.

Several areas factor into the decision on the best acquisition path for the Future Frigate. Table 3.1 discusses the issues in these areas for the three broad acquisition paths.

Important Initial Program Questions

As the program moves forward, several near-term questions should be addressed. These include the following:

- *Can an existing ship design meet the desired Future Frigate's operational requirements?* The answer here could eliminate the pure MOTS option.
- *If no existing ship design can meet desired operational requirements, what modifications to the basic design are needed?* Existing designs may require only minor modifications or could be quite substantial.
- *Will the desired schedule for the fleet introduction of the Future Frigate preclude any design and construction options?* The answer to this question could eliminate a new design option.
- *What are the total ROM life-cycle costs of each feasible acquisition path for the Future Frigate?*
- *What is the trade-off between the operational capabilities of the ship and the resulting costs, schedule, and risks of the different acquisition options?*

Table 3.1
Summary of Acquisition Options

Factor/Element		Pure MOTS	Evolved MOTS	New Design
Requirements		Existing designs may not meet desired capabilities	Can shape existing designs to meet desired capabilities, but limits may exist	Can design ship to meet desired capabilities
Cost	Design	Minimal	Depends on extent of modifications	High
	Construction	Depends on construction strategy and proficiency of construction shipyards		
	In-service	Depends on complexity of design, availability of technical data rights, and proficiency of support organizations		
Schedule		Minimal design, construction, and test time	Depends on extent of modifications	Potentially the longest schedule
Required resources	Design	Minimal	Depends on extent of modifications	High
	Construction	Depends on construction strategy and proficiency of construction shipyards		
	In-service	Depends on complexity of design, availability of technical data rights, and proficiency of support organizations		
Risk		Low with proven design	Depends on extent of modifications	High depending on deviation from existing designs
Competition		High among feasible existing options	High depending on industrial base strategy	Could be high among partner design organizations, shipbuilders, and equipment suppliers

- *What technical data rights and approvals are needed to adequately upkeep, update, and upgrade the capabilities of the Future Frigate over the life of the class?* Lessons from previous and current programs can help specify exactly what technical data rights are needed.

CHAPTER FOUR

Overview of a Naval Shipbuilding Program

This chapter provides a brief overview of a major ship acquisition program. It first describes how ship programs differ from other defense projects and then lays out the major phases of a naval shipbuilding program. The chapter then describes the two major processes used for the design and construction of a ship: a sequential design and build and a concurrent design and build. The chapter also includes definitions of the significant decision milestones of a major acquisition program and the roles and responsibilities of the organizations, both public and private, that are part of a program.

How Are Ships Different from Other Weapon System Acquisitions?

Naval vessels represent some of the most complex and challenging defense acquisition programs that a country can undertake. Warships integrate multiple weapon systems and can host other smaller vehicles, such as unmanned aerial vehicles (UAVs), helicopters, and small boats. Beyond their direct warfighting roles, warships must accommodate (e.g., berth, provide office space, feed, and supply) tens if not hundreds of crew for extended periods of time. They must be able to deploy and function for months at a time. Most important, the very first ship produced in a class must be a functioning asset. There is no prototyping of ships; to do so would be a prohibitively expensive practice.

Hence, ships are much more complex than other weapon systems. In 2002, General Dynamics Electric Boat compared the characteristics of three complex systems. We reproduce this analysis in Table 4.1.

This complexity is just one distinguishing aspect of naval ships. In prior research,[1] RAND interviewed numerous stakeholders in U.S. naval ship acquisition. These interviewees identified aspects that are different for naval ships than for other weapon systems:

- greater time to design and build (this also applies to extra time needed for long-lead items)
- greater influence of political factors (shipbuilders tend to be large regional employers)
- high concurrency of design and build (there is significant overlap between design and construction on the first of class)

Table 4.1
Comparison of Technical Characteristics of Three Programs

	Virginia-Class Submarine	Boeing 777 Aircraft	M-1 Tank
Weight (tons)	7,800	250	65
Length (feet)	377	200	25
Number of systems	200	40	25
Patrol/sortie duration (hours)	2,000	8–14	24
Crew size	113	10	4
Unit production time (months)	55	14	7.5
No. part numbers	1,000,000	100,000	14,000
No. manhours/unit	>10,000,000	50,000	5,500
Annual production rate	0.5–2.0	72	600

SOURCE: General Dynamics Electric Boat, *The VIRGINIA Class Submarine Program: A Case Study,* Groton, Conn., February 2002.

---

[1] Jeffrey A. Drezner, Mark V. Arena, Megan P. McKernan, Robert E. Murphy, and Jessie Riposo, *Are Ships Different? Policies and Procedures for the Acquisition of Ship Programs,* Santa Monica, Calif.: RAND Corporation, Santa Monica CA, MG-991-OSD/NAVY, 2011.

- higher complexity (more systems, greater human factors issues, greater systems engineering challenges)
- low quantities and production rates
- higher unit costs
- different test and evaluation approaches (e.g., there is no full-system prototype, testing cannot damage the ship, often testing is not complete on first hull when production on following hulls begins).

Because of all the differences described above, ships tend to have a unique acquisition process.

Ship Program Phases[2]

There are eight distinct phases to the naval ship program life cycle. These phases are as follows:

1. Solutions Analysis: This phase is a broad exploration of materiel solutions that may meet requirements to fill identified military capability gaps.[3] It should be led by the government but may be assisted by industry. Multiple alternatives are explored to understand the cost-effectiveness of various system choices.[4] Part of the activity during this phase is to examine requirements and determine if they are affordable. At the end of this phase, the government selects a single concept to refine in the next phase.
2. Concept Design: During this phase, various military requirements are traded against ship size and cost. The output of the phase should be a conceptual design (basic arrangements, system descriptions, high-level work-breakdown structure, etc.)

---

[2] This section was adapted from Naval Sea Systems Command (NAVSEA), *Description of the Naval Ship Design Phases,* briefing, Washington, D.C., February 2008.

[3] *Requirements* in this document mean the mission capabilities and not the performance characteristics of the system itself.

[4] By *system* in this context, we mean some major element of the ship, such as propulsion plant, sensors, a weapon, or aviation support equipment.

and a preliminary weight statement (weights for each element of a high-level work-breakdown structure). Another important output of this phase is the identification of major risks and a plan to mitigate them. Mitigation may include systems development, component testing, and parallel technology developments. Last, a number of engineering reports should document the concept design iterations and trade-off studies performed.

3. Preliminary Design: The purpose of this phase is to fully define the ship characteristics, establish system architectures, and establish a detailed cost baseline using the conceptual design output from the prior phase. During this phase, major components, hull form, and new technologies are selected. The output for this phase includes a general arrangement drawing, systems diagrams, a list of major equipment, a detailed weight estimate, a budget-quality cost estimate, a detailed key event schedule, and a detailed, resource-loaded work-breakdown structure.
4. Contract Design: Contract design has traditionally been the engineering development of the technical and contractual definition of the ship design (including ship specifications and drawings) to a level of detail sufficient for shipbuilders to make a sound estimate of the construction cost and schedule. New technology developments are typically initiated during this phase. There are numerous technical products; some of the most significant are definition of major interfaces and configuration control plans. Also, a build strategy is usually finalized during this phase.
5. Detailed Design and Construction: These activities are highly concurrent irrespective of the design and build strategy discussed below, so we will describe them together as a single phase. During this phase, the ship design is fully defined in terms of product models, construction drawings, and procurement specifications for material, equipment, and systems. Construction begins as design products are finished. Logistic support plans and crew training materials are also developed during this phase.

6. Test and Trials: As hulls are delivered, they go through an evaluation phase where all the operational aspects of the ship are checked. This process is typically done with a combination of Navy and contractor personnel. The test and trials are much more extensive for the first hull to prove the design. Once the test and trials are finished, the ship returns for a brief refit and repair period to address any problems. Also, certain systems of the ship may be upgraded at this point, since they have not changed since the beginning of construction (which may have begun several years earlier). In execution, testing begins during the construction phase.
7. Operations and Support: Once the test and evaluation phase is finished and discrepancies are addressed, the ship enters the operations and support phase. The program office still has a role in planning and executing maintenance as well as improvements and upgrades. Depending on the support strategy selected, the builder/designer might have a significant role, as well.
8. Retirement/Disposal: Once the ship has reached the end of its useful life, it is either sold or scrapped according to the laws of the country. Scrapping a ship can be an involved activity given various environmental laws.[5]

All of the above phases apply regardless of the acquisition path although to various degrees. For a pure MOTS or evolved MOTS option, the solutions analysis identifies and evaluates various existing designs and chooses one for further exploration. The concept, preliminary, and contract design phases are basically complete for a pure MOTS acquisition path. An evolved MOTS acquisition path will require some degree of concept, preliminary, and contract design, with the degree of involvement dependent on the magnitude of change from the baseline ship design. A new design will need to progress through

[5] We do not discuss this phase further in this report. Interested readers should see Ronald Wayne Hess, Denis Rushworth, Michael Hynes, and John E. Peters, *Disposal Options for Ships*, Santa Monica, Calif.: RAND Corporation, MR-1377-NAVY, 2001.

these three design phases completely. The last three phases are needed whether a MOTS, evolved MOTS, or new design is chosen.

In prior work for the Australian Department of Defence (DoD), RAND conducted an overview of the submarine design process, which is the same as that used in surface warship design (although the content and technical issues are quite different). We reproduce key aspects of this discussion in the next section.

Design-Build Sequence[6]

Ship design is a transition through the four design phases (contained within the eight program phases listed above) whereby the ship design increases in fidelity and complexity. The process culminates in a package of drawings supporting the shipyard construction and sufficient information allowing component procurement from subcontractors. The traditional design process goes sequentially through each of the design phases and requires that one phase be completely finished before the other begins. In modern projects that employ design-build methods, these phases are highly concurrent. Thus, where one phase begins and ends is not as clear.

Sequential Design-Build

As shown in Figure 4.1, the traditional sequential process views each phase of design as distinct, ending before the start of the next design phase.

Traditionally, the four phases of the design process (concept, preliminary, contract, and detail) are conducted in a lock-step manner, with a period between each phase where decisions are made on whether, and how, to proceed with the overall design program. The rationale behind sequential design is to divide large complex efforts into discrete,

---

[6] This section is adapted from John Birkler, John F. Schank, Jessie Riposo, Mark V. Arena, Robert W. Button, Paul DeLuca, James Dullea, James G. Kallimani, John Leadmon, Gordon T. Lee, Brian McInnis, Robert Murphy, Joel B. Predd, and Raymond H. Williams, *Australia's Submarine Design Capabilities and Capacities: Challenges and Options for the Future Submarine*, Santa Monica, Calif.: RAND Corporation, MG-1033-AUS, 2011.

Figure 4.1
Sequential Design-Build Process

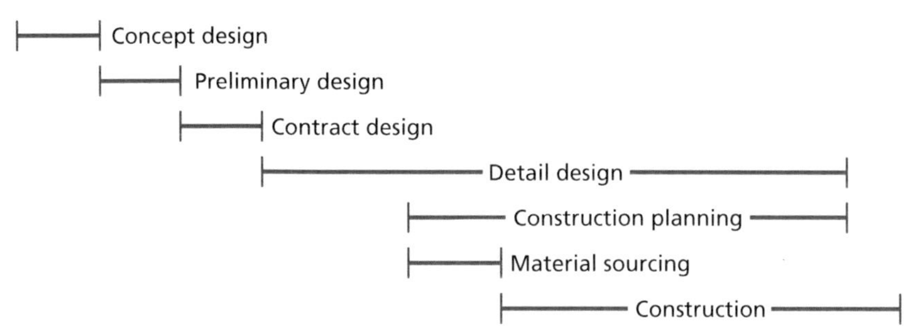

SOURCE: Birkler et al., 2011.
RAND *RR767-4.1*

manageable efforts. This approach has the benefit of providing time for the government to make decisions before a program advances to its next development stage.

However, one drawback of this approach can be continuous design revisions. The process begins with designers proposing a design solution. Next, the procurement group issues subcontracts for the design or manufacture of required material and components. The feedback from component developers often indicates performance shortfalls, and new component development is proposed or the ship design is modified to take advantage of alternative components or technology. Logistics personnel then perform a review. Once again, a series of negotiations takes place and design revisions and re-approvals occur. At some point, the end user is invited to see the product and provide expertise and comments on the man-machine interface, human system integration, and life-cycle maintenance. Once again, comments are generated and the design is revised. Finally, production personnel review the design and a further series of negotiations takes place to modify the design to facilitate production.

As fidelity increases, additional reviewers address other areas—environmental issues, training, safety, and the like. At each step, the potential to discover a new flaw exists, the resolution of which requires redesign. Complicating the effort, external technical reviewing authorities often require mutually exclusive design solutions. Because they have no ownership of the design, these reviewing authorities have no incentive to compromise or accept risk and view themselves as gatekeepers of technical purity. Consequently, delays and disruptions may arise as multiple layers of senior decisionmakers become involved in resolving issues.

Significant drivers behind higher risks and costs are the multiple redesigns of completed ship sections—motivated by the need either to (1) accommodate requirements changes initiated late in the design by the customer, the technical authority, or the manufacturer or to (2) address flaws that need to be resolved at the design contractor level. The later in the process these events occur, the more costly the resolutions, as they affect ever more detailed drawings. Design flaws found during construction and test are the most costly to correct.

A second trait of the sequential design process is the need to complete each phase in sequence before starting the following phase. For example, the contract design tasks do not start until all preliminary design efforts are completed. This approach generally stops work in selected areas to wait for a critical path to complete, which can lead to staffing issues for specific skill sets and the arbitrary lengthening of the design process.

These clearly defined stopping points can be advantageous to customers, allowing them to perform a holistic review of the design at some known level of fidelity. However, these intermediate intervals between design phases can delay the design process, disrupt the workforce, and often result in changes to requirements or preferred approaches to a design solution. Any such changes become increasingly disruptive and costly as the design stages progress.

Moreover, design products for use in a competitive award of a subsequent design or construction contract will be generic and not optimized for any one contractor. Intellectual property and competitive advantage concerns also may keep a set of designers from disclosing their best ideas to the customer in these pre-competition design products. Similarly, intellectual property concerns may prevent collaboration between prospective competitors, reducing the pool of skilled resources available to work on early stages of the design process.[7]

Table 4.2 summarizes some advantages and disadvantages of sequential design.

Concurrent Design-Build

To overcome some of the limitations of the sequential design process, some organizations have adopted concurrent engineering processes, wherein discrete, sequential design phases overlap. That is, the last three phases of the design process—preliminary, contract, and detail—are performed in a seamless manner without the start and stop seen in sequential design. This concurrent process may be inherently more manpower- and resource-intensive, since more design work is being performed at any given time in the phases of design. The risk is that

Table 4.2
Advantages and Disadvantages of a Sequential Design Process

Advantages	Disadvantages
Clearly defined review points	May result in longer design times
Workforce demands more evenly spread with time	Prone to greater levels of rework
Clear organizational responsibility during each phase	"Throw it over the fence" mentality between designers, suppliers, builders, and customer
Clear points to hold competitions	Minimal participation from manufacturing, operations, test, and support communities
	Difficult to keep workload uniform over design

SOURCE: Birkler et al., 2011.

[7] In the United States, the *Los Angeles*-class submarines were noncompetitively designed using the sequential design process. The *Seawolf*-class submarines were competitively designed using the sequential process. The *Virginia*-class submarines were noncompetitively designed using a concurrent design process. Although the *Los Angeles*-class submarines were not as capable or complicated as the other two classes, they were designed and constructed in the shortest period of time—approximately seven years shorter than either of the other two classes.

problems may surface only after a great deal of engineering effort has been expended.[8]

To overcome this problem, using the concurrent design approach requires more collaborative design teams that include production and the owner. One advantage of such an approach is that problems often surface earlier and are resolved when it is less expensive to make changes. This collaborative, concurrent design process is better known as Integrated Product and Process Development (IPPD). The IPPD process merges many stakeholders earlier into the design process than the sequential process. It balances the complexity, breadth of technical skills required, duration of design effort, and cost sensitivity to schedule disruption for extremely complex aircraft and ship programs. The approach begins with the end in mind and folds in all aspects of the product life cycle (e.g., production, procurement, test and evaluation, support, etc.). It considers more than just the design outputs; it also takes into account such issues as manufacturability and supportability, thereby reducing changes later in the design, build, and maintenance of a product.

IPPD processes have demonstrated significant improvements/efficiencies in the delivery of complex multidisciplined products within cost and schedule constraints. Their successes include programs ranging from the Boeing 777 and F-18 E/F fighter acquisition to the *Virginia*-class submarine produced by General Dynamics Electric Boat.[9]

General Dynamics Electric Boat's experience is instructive. In 2002, it reported "Problems identified during construction are far fewer and less serious for *Virginia* than *Seawolf* . . . as of the end of Jan-

---

[8] Three cost axioms of ship design are (1) most of the future construction and maintenance costs to be incurred are locked into the design in the early part of the effort, (2) the cost to make a change is lowest in the early part of the design effort and increases proportionally as the design matures, and (3) changes made during construction are the most costly.

[9] Boeing, for example, cites a reduction in cycle-time of 17 percent and a reduction of rework of 40 percent. See Gary Brown and Cliff Harris, "Matching Product Development Practices to the Product Life Cycle," Center for the Management of Technological and Organizational Change (CMTOC), *Highlights of the Thirty-Fifth Advanced Manufacturing Forum*, February 27–March 1, 1995.

uary 2002, 3.2 years after construction start, the *Virginia* builders had identified about 5,300 problems. As a fraction of labor hours required to build the ship, *Virginia* had reached almost 70 percent. *Seawolf* did not reach that level of construction completeness until almost six years into the build. At that time, *Seawolf's* builders had identified about 53,700 problems. So, the reduction in errors at comparable points of completion is about 90 percent."[10]

General Dynamics Electric Boat also showed a negligible growth in manhours at completion for *Virginia* compared to 77 percent for *Seawolf* and 60 percent for Ohio.[11] However, in its 2002 report, General Dynamics Electric Boat acknowledged that the experience gained in designing and building *Seawolf*, in addition to the IPPD approach, contributed to many of the improvements in design and construction efficiency it saw on *Virginia*. For example, the heavy computational analysis and testing performed on *Seawolf* to improve its acoustic signature were rolled over into *Virginia*. In another example, significant problems with technical welding on the *Seawolf's* pressure hull were resolved over the course of several years, such that they did not occur at all on *Virginia*. With any luck, experience from the AWD program will similarly result in design and construction efficiencies for the Future Frigate.

A concurrent design approach using IPPD starts with a systems definition phase followed by an integrated design/construction planning development phase, as shown notionally in Figure 4.2.

An important aspect of the concurrent design approach using IPPD is the use of a design/build/support approach. This philosophy integrates individuals who are knowledgeable about the construction and in-service support processes into the design teams very early in the design activity. Bringing construction and in-service support expertise to bear early in the design process can minimize costly rework during construction that stems from a mismatch between what designers desire and what builders and maintainers can efficiently build and support. The result is far fewer design changes during construction.

---

[10] See General Dynamics Electric Boat, 2002, p. 69.

[11] General Dynamics Electric Boat, 2002, p. 17.

Figure 4.2
The Concurrent Design Process with IPPD

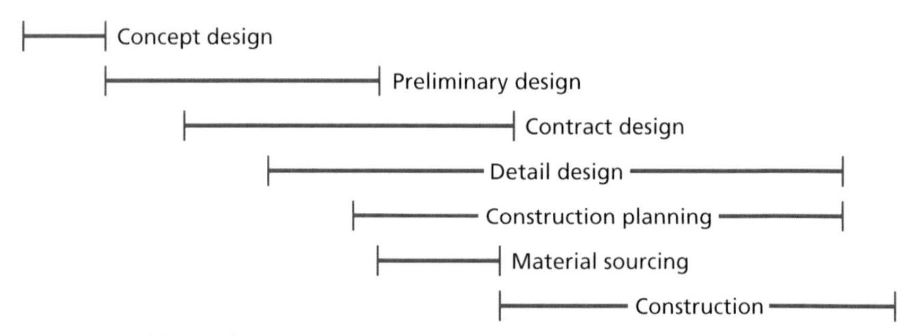

SOURCE: Birkler et al., 2011.
RAND *RR767-4.2*

Design/build/support is, at its simplest, an industry-driven, system-engineering process established to deliver a product. It encompasses a design philosophy that is driven by integrated, multidiscipline teams, preferably co-located, wholly accountable for the cost and technical quality of the product.

Table 4.3 summarizes some advantages and disadvantages of the concurrent design approach using IPPD.

In terms of the three options the SEA 5000 program is considering, the pure MOTS option will need to significantly alter the design-build process. Because the detailed design will be complete, the design phase can start somewhere in the production planning process. If an overseas build option is selected, the production drawings and instructions should exist; the program can directly proceed to material sourcing and construction. If a domestic build option is chosen, the program will need to generate new construction drawings and plans that will be consistent with the resources and infrastructure at the building yard(s). With the evolved MOTS option, the program might proceed quickly through the concept, preliminary, and detailed design, if there are minimal changes. More significant changes will require more time. The new design option will require the most time and will need to go through all the phases.

Table 4.3
Advantages and Disadvantages of Concurrent Design Process That Includes IPPD Processes

Advantages	Disadvantages
Shorter design cycle-times	Lack of clear review points
Less design rework	Highly concentrated workforce demands
Collaborative process encompassing several stakeholders	Challenging program management
Better manufacturability	Difficult to have production competition after collaboration in design
Potential to decrease lead ship and recurring construction costs	Need for co-located teams
Potential to decrease maintenance costs	Government must provide timely input
	Increases up-front nonrecurring design costs; the design funding profile must be front-end loaded
	Requirement to pick builder at same time as designer—limiting competitive design options

SOURCE: Birkler et al., 2011.

Program Oversight: Major Decision Milestones

Given the complexity of ship programs and the variety in the design and build processes, there is no universally accepted approach to milestones and oversight. Even in the U.S. context, where there is a well-defined acquisition process and milestones, ship program oversight must be highly tailored to be useful and fit within the system[12] and this is equally true in Australia.

However, we can think of a set of generic milestones and oversight activities from which to develop a detailed oversight process. *The generic milestones are tied to major decisions by the government on the release of funds or contract actions.* The entire process begins with a decision that there is a valid need for a new system to fill an identified or established gap. Usually this decision is accompanied by a set of operational

[12] Drezner et al., 2011.

requirements. With an established gap and requirements, conceptual design work begins that seeks to understand the range of options that can meet those operational requirements and their technical feasibility along with cost. In the Australian context, this phase results from First Pass approval from government. As a result of the Mortimer report of 2008 reforms that established this approval process, it can be expected that up to 15 percent of the estimated total program cost can be used to support the phase(s) between First and Second Passes.[13]

The first real program milestone is then the selection of a concept (i.e., a materiel solution) to further refine. This selection begins the design-build phase of the program. Formally, with Second Pass approval from government, the full release of funding for the project is obtained when the design is mature enough to have confidence in the program's cost. Generally, this stage of maturity enables a detailed design and construction contract to be negotiated. However, as indicated above, shipbuilding is a complex process, and so it is likely that there will be multiple acquisition Passes by government as the design and construction progresses.

The process radically changes depending on the design-build sequence. *With a sequential design process*, each design phase culminates in a review of the technical products from that phase and a decision to proceed. Once the detailed design has reached sufficient maturity, a decision is made to begin construction. *With the IPPD process*, review by the customer is almost continual during the design phase. At some point, design products reach sufficient maturity and are released to production. This process continues until all design products are done and the ship construction is finished.

Both processes re-emerge onto a similar path at the completion of construction. The newly completed ship goes through test and trial to make certain that it fulfills all the requirements and functions adequately. At the end of this period, the ship typically undergoes a brief refit period to address defects and deficiencies. This occurs for all ships,

---

[13] David Mortimer, *Going to the Next Level: The Report of the Defence Procurement and Sustainment Review*, Defense Materiel Organization, 2008.

but at the end of the trials for the first ship of the class a decision is made for additional production.

In Figure 4.3, we show a generic *acquisition* process with four distinct acquisition phases[14] overlaid with major milestones in the shipbuilding design and construction process for the first hull.[15] At the top of the figure are the shipbuilding phases (discussed above) overlaid on this same timeline. Also included in Figure 4.3 are the notional indications of First and Second Pass approval for the Australian acquisition process. However, as mentioned above, in most Australian shipbuilding programs, multiple passes through government are likely because of both the large size of a shipbuilding program and its political and economic implications. Notice that the shipbuilding phases do not cleanly map to the acquisition phases. Thus, one challenge of a shipbuilding program is tailoring the acquisition phases (generally designed for large production runs) to the shipbuilding phases.

The solutions analysis phase is the first acquisition phase. This phase explores the possible materiel solutions to meet the mission requirements. It is similar to the first ship program phase, but some conceptual design may begin during this acquisition phase. The concept refinement phase combines conceptual, preliminary, and contract design activities (Shipbuilding Program Phases 2 through 4). The goal of this acquisition phase is to further define the selected solution and reduce risk. The next acquisition phase is detailed design and build (shown together as they tend to be concurrent for ship programs)—similar to the Shipbuilding Program Phase 5. The last acquisition phase is test and trials phase, wherein the first hull is tested and evaluated in terms of the requirements (Shipbuilding Program Phase 6). Note that some early testing begins during the detailed design and build phase (e.g., material receipt and inspection); but to simplify the diagram, we

---

[14] Note that these phases are fewer than the eight shipbuilding program phases introduced above. Nor do they match exactly in name. Most acquisition systems review programs only at major milestones, which are tied to either significant technical products, release of funds, or contract actions.

[15] Other hulls in the class will repeat the build, test, and trial phases (although the trials may not be as extensive as for the first hull). If the class is developed in flights, then some of the design activity may be repeated, as well.

Figure 4.3
Shipbuilding and Acquisition Phases, Decision and Requirements Milestones, and Australian Passes for First Hull

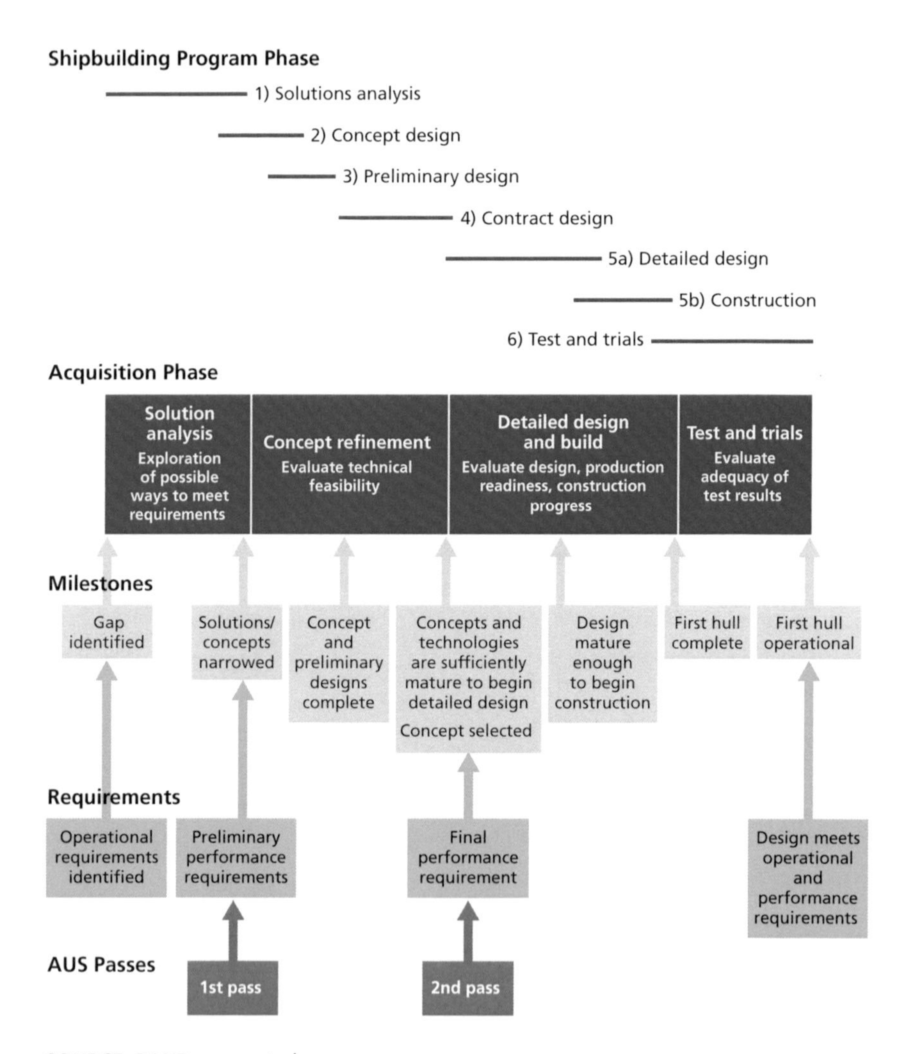

SOURCE: RAND-generated.
NOTE: Phase durations are not to scale. Sizes of phases were chosen to simplify the display to show overlaps of activities.
RAND *RR767-4.3*

have shown it as a distinct phase. The actual trials begin once construction is complete.

The typical decision and requirements milestones are also shown in Figure 4.3. These milestones are not a complete list, but rather highlight key decision points in the process. The decision milestones are:

1. Gap identified (begins solutions phase) and some conceptual design begins.
2. Solution (concept) is selected/ready to start conceptual design refinement.
3. Concept and preliminary design are complete. Technology risk reduction begins (although some prototyping efforts may have started).
4. Technologies are mature and contract design complete. Contract award for detailed design may happen.
5. Detailed design is sufficiently mature that construction may begin (along with contract award for production of first few vessels).
6. Delivery of first hull; test and trials phase begins.
7. The conclusion of the trial phase ends with the first hull entering service. Decision to continue production (not shown—can occur anywhere in the design-build phase or test and trials).
8. First hull is operational.

The requirements milestones are:

1. Operational Requirements: These are identified.
2. Preliminary Performance Requirements: For the Australian acquisition system, these requirements derive from identified measures of effectiveness (MOE) and measures of performance (MOP) that are prioritized into levels: essential, important, and desirable. These requirements are documented in the Preliminary Functional and Performance Specification. The U.S. acquisition system identifies analogous requirements as key performance parameters (KPPs) and key system attri-

butes (KSAs)—although the preliminary values are determined slightly later in the process (at the end of preliminary design).

3. Final Performance Requirements: For the Australian acquisition system, the final threshold and objective requirements are determined, with particular emphasis on the essential and important requirements. For the U.S. system, these requirements would correspond to the final KPPs and KSAs.
4. Product Evaluation: The product is evaluated to meet the stated requirements and determine if all deficiencies have been addressed (and design is updated).

General Timeline of the Process

As with the timing of milestones and the structure of the process itself, the duration of the steps can vary widely. In Table 4.4 we summarize notional durations for acquisition phases.

The ranges in Table 4.4 reflect a complex interaction between industrial capabilities and the product (ship complexity). More complex or larger vessels will require more time. Similarly, more introductions of new technology will lengthen the concept refinement phase. As discussed above, the time to go through these phases also will depend on the option chosen. A pure MOTS option may have very short preliminary and contract durations, for example.

Table 4.4
Notional Durations of Acquisition Phases for Naval Ships

Phase	Duration (years)
Solutions analysis	1 to 3
Concept refinement	2 to 4
Design and build	4 to 7
Testing and trials	1 to 2

SOURCE: Adapted from NAVSEA, 2008.

In Table 4.5, we present data and timelines for recent U.S. and UK ship designs. We also show cost and quantity data for the programs so that the reader may put each program in context in terms of their relative scopes.

Ship Construction and Testing Milestones[16]

As with the design-build oversight process, several key events occur during construction and testing (Shipbuilding Program Phases 5 and 6) that are not fully shown on Figure 4.3. Some of these events coincide with oversight (such as ship acceptance) whereas others are tied to specific construction events.

- Advanced Appropriation: Often funding is required before formal authorization to procure long-lead items for construction. Without such early purchases, the ship delivery could be greatly extended or delayed while waiting for such items. Also, long-lead delivery allows for the timely installation of systems so that they are installed on the ship at the most cost-effective time during construction.
- Ship Authorization: Full funding is approved for new construction (Second Pass in the Australian context).
- Keel Laying: This milestone is the formal recognition of the start of a ship's construction. Historically, ships were constructed from the keel upward. So the keel laying milestone marked the start of major construction activities. Today, fabrication of the ship may begin months before and some of the ship's hull sections may already be joined before this event. However, keel laying symbolically recognizes the joining of modular components in the dock and the ceremonial beginning of a ship.
- Launching: This is the point when the ship enters the water for the first time. Traditionally, it coincided with the ship's christe-

[16] These definitions are adapted from Navy League of the United States, "Shipbuilding Milestones," undated. We have omitted the ceremonial milestones, such as Christening and Ship Naming.

Table 4.5
Comparison of Select Ship Program Characteristics

		Ships					
Difference	**Metric**	**DDG-51**	**LPD-17**	**SSN-774**	**DDG-1000**	**CVN-78**	**Type 45**
Program initiation[a]	Start of concept refinement	February 1980	November 1990	August 1992	June 1995	March 1996	[b]
Unit cost	PAUC	$1,084 ($M, BY 2006)	$1,352 ($M, BY 2006)	$2,536 ($M, BY 2006)	$3,659 ($M, BY 2006)	$9,307 ($M, BY 2006)	£944 (£M, TY)
R&D funding	% RDT&E funding of total funding	6%	1%	8%	35%	14%	[b]
Quantity	Planned quantity	62	9	30	7	3	12
Production rate	LRIP quantity	9	No LRIP	18	7	3	3
	Annual rate (min/max)	1/5	0/2	1/2	1/1	0/1	0/1
Production strategy	Allocation of production work	Dual shipyards, whole ships	Single source	Dual shipyards—modules	Dual shipyards—modules	Single source	Dual shipyards—modules
Design phase time	Conceptual design to detailed design (months)	30	42	11	96	47	[b]
	Detailed design to first delivery/acceptance (months)	91	110	113	89	141	96
Size	Full displacement (long tons)	9,515	2,5883	7,008	15,656	112,000	7,900
	Overall length (feet)	510	684	377	610	1,092	500
Crew	Accomm. (A) or crew (C)	312 (A)	396 (A)	132 (A)	158 (C)	4539 (A)	190 (C)

SOURCES: Drezner et al., 2011; National Audit Office, Ministry of Defence, *The Major Projects Report 2011: Appendices and Project Summary Sheets*, HC 1520-I, Session 2010–2012, November 16, 2011; NAVSEA Shipbuilding Support Office, *Naval Vessel Register: Inventory of US Naval Ships and Service Craft*, NVR Online, updated October 6, 2014.

[a] Program initiation is officially later for U.S. programs. Here, we mean the first milestone in which a coherent program is presented to decisionmakers (which is earlier).

[b] Comparable data do not exist, as the program evolved out of two prior programs that were cancelled.

ing, with the ship sliding down the ways into the water with a splash. Today, many launchings take place separately from the christening—ships are floated away from a dry dock or a lift facility. Typically, the ship is then pulled alongside a pier for final outfitting and finishing. The degree of completeness for the ship at launching depends on the ability of the facilities to lift or launch ships—they are usually weight- or draft-limited.

- Testing: Testing begins during the construction phase with the inspection of material and equipment. The testing then proceeds into installation inspections and equipment tests. After launch, major system checks begin to ensure that entire systems behave as designed. Tests of multiple systems together are conducted to ensure that systems interact and operate together (or do not interfere with one another).
- Sea Trials: These involve an intense underway period to demonstrate the satisfactory operation of all installed shipboard equipment and performance of the ship as a whole in accordance with the plans and specifications. New construction ships undergo builder's trials and acceptance trials before ship's delivery. Final contract trials occur several months *after* delivery and sail-away.
- Contractor Fitting Out (CFO) Date: The major parts of CFO include such activities as inspecting, staging, inventory accuracy, loading of authorized material, and identification of requirements.
- Delivery: The official turnover of custody of a ship from the shipyard to the Navy. This event normally coincides with "Move Aboard" when the precommissioning crew moves aboard and starts living, eating, standing watch, training, and working aboard the ship while final work continues in the shipyard.
- Post Shakedown Availability (PSA): The PSA is an industrial activity to correct deficiencies found during the shakedown cruise or to accomplish other authorized improvements. PSAs are scheduled to commence after delivery, typically 6 to 12 months later. The PSA is also an opportunity to install updated systems and software (typically weapon systems and C4I systems).

An acquisition and contracting strategy encompasses the various steps and milestones described above. We discuss this in greater detail in the next chapter.

CHAPTER FIVE

Acquisition and Contracting Strategy

Establishing a transparent and fair acquisition and contract environment can set the tone for any new program. Successful programs have strong and clearly defined partnerships built between the program office, prime contractors, and subcontractors. In this chapter, we discuss various forms of acquisition and contracting strategies and issues around them that have had an effect on recent shipbuilding programs. We also discuss best practices and key lessons that may inform the planning phases of the Future Frigate program.

Government Interaction with Industry

Part of the acquisition strategy defines how the government interacts with industry. The classic approach is through a single prime contractor that manages the detailed design and build of most aspects of the program (with the possible exception of government furnished equipment [GFE]). Such an approach works best when industry (through a prime contractor) has extensive capabilities to control all of design and construction. This is the predominate model in the United States.

Where the prime does not have the ability to fully control the process (or does not own such a capability), a few alternative models are possible. In an alliance model, currently used for both the AWD in Australia and the *Queen Elizabeth* aircraft carrier (CVF) program in the UK, government and commercial companies form partnerships. Such an approach is generally chosen where there is high risk.[1]

In the UK case, no single industry partner had the resources to fully execute the program. For the AWD case, the program was viewed as high risk because of the long gap in naval ship production and the need to align the designer, builder, and weapon systems provider. Also, the alliance approach is used in the commercial realm for similar high-risk reasons (and fixed-price contracts are too expensive or unattractive to industry) or when schedule is a priority.[2]

Engineer, procure, and construction management (EPCM) are common commercial approaches in the capital investment world (e.g., oil and gas). For such a contract, the prime will be responsible for design and construction management but will subcontract to local construction firms for the execution of the construction content.

This potential government-industry structure will be highly dependent on the technical solution chosen. For example, with a clean-sheet design, a prime or EPCM approach would be possible (given adequate domestic resources). For a MOTS solution, the options will depend, in part, on the chosen acquisition strategy for the build location. If a domestic build option is chosen, an alliance might be the preferred approach, as it would help to integrate the design owner and the build firm(s).[3] Otherwise, government will have to act as the intermediary between the two organizations and that would require substantial management resources and risk ownership by the government.

The roles and responsibilities of the government and industry evolve over the course of a program. Typically, in the earlier phases of design, the government has a stronger or lead role. Leadership eventually transitions to industry once the demands for technical resources grow to a point that the government cannot lead design activities. Almost universally, industry leads the production phase. The timing of

---

[1] See, for example, John Paul Davies, *Alliance Contracts and Public Sector Governance,* Ph.D. thesis, Griffith Law School, Griffith University, South East Queensland, Australia, August 2008.

[2] Davies, 2008.

[3] Another advantage of an alliance is that it helps to lock in firms early.

the transition from government-led to industry-led (if it happens at all) depends upon the following three factors:

- Resources and skills of each party: A government program office with deep technical skills and resources typically leads the earlier design phases with industry in support. Later, the lead will switch to industry. Where the government does not have the depth of knowledge or expertise, it must supplement its skills with outside support or contract/team with industry directly. For an established system, government might award a sole-source contract from the feasibility point and let industry translate requirements into conceptual designs.
- Acquisition strategy: How the government chooses to buy a system can also greatly influence who has the lead and when. If the procurement will be competitive during the design phase, industry teams might lead individual design efforts. The government will follow the teams (but not direct them). For its newest Offshore Patrol Cutter, for example, the U.S. Coast Guard is using a competitive strategy for both preliminary and contractual designs; this strategy carries multiple teams up to the detailed design and construction award. For an uncompetitive award (sole source), the government might lead the initial design phases and then work collaboratively with industry up to contract design. The AWD program adopted an alliance acquisition approach to share risk and achieve alignment between the various industry partners.
- Design maturity of the concept: The extent to which design information already exists when the concept is selected will influence the roles in the early design phases. If a MOTS solution is chosen, then the industry owner of that design will have the lead for all the design activity. For a clean-sheet (new) design, the design activity could be led by industry or the DMO with RAN involvement.

In Table 5.1, we show some acquisition strategies[4] and the design and construction lead role. Note that we do not show all possible options, only a few of the more common ones.

Different Types of Contracts

Military shipbuilding is typically characterized by low volume and limited sources, both of which restrict the government's ability to pursue competitive acquisition strategies that it might normally select for other types of procurements. In some cases, the government may

Table 5.1
Example Acquisition Strategies and Organization in Lead Role, by Design-Build Phase

Acquisition Strategy	Concept Refinement	Preliminary Design	Contract Design	Detailed Design and Construction	Test and Trials
Government design	Government	Government	Government	Industry	Joint
Industry design (sole-source—prime)	Government	Industry	Industry	Industry	Joint
Competitive design (down-select to single prime)	Government or industry (multiple firms)	Industry (multiple firms)	Industry (multiple firms)	Industry (single firm)	Joint
MOTS (turn-key)	Industry	Industry	Industry	Industry	Joint
Alliance	Collaborative	Collaborative	Collaborative	Collaborative	Joint

SOURCE: RAND-generated.

NOTES: Within the test and trials phase, the government and industry have dual responsibilities depending on the specific activity being demonstrated or system or equipment being tested. The government bears responsibility for performance of GFE and must assure itself that contractor furnished equipment (CFE) and systems operate properly. The contractor must demonstrate the proper operation and integration of equipment and systems it installed.

[4] By *acquisition strategy*, we mean the way the government chooses to interface with industry and execute the program.

be the shipyard's sole customer. Because of this largely monopolistic relationship, there is a greater need to carefully design contracts with incentive structures for cost, schedule, and performance. In addition, shipbuilding programs often include a number of subcontractors across various systems. Relationships between these suppliers that are not clearly defined can degrade integration and performance during the design and build phases of the program.

Two broad categories of contracts for government procurement are firm fixed-price (or fixed-price) and cost-reimbursement (or cost-plus) contracts. The former typically places more risk on the contractor to ensure performance under a firm ceiling price or target price with profit incentives (target cost incentive). Firm fixed-price contracts tend to be more effective in situations where there are well-defined ship specifications and reliable cost estimates, such as when procuring military or commercially proven platforms. However, in developmental projects where high levels of cost uncertainty exist, firm-price contracts may lead to failure if unforeseen cost overruns exceed the contractors' ability to absorb those costs. Ultimately the government will still have to absorb those cost overruns (even those above the ceiling) or face program cancellation.

It is possible that a number of contracts will need to be negotiated for the SEA 5000 program; for example, design contracts, acquisition or construction of the first of class, construction of follow-on ships, and life-cycle support contracts. In addition, the preferred contract type may vary by the acquisition strategy. Under a pure MOTS acquisition strategy for the Future Frigate, given that the design and technology of the platform is proven, the best contract vehicle for the construction of all ships of the class might be firm fixed-price. However, in an evolved MOTS acquisition strategy where significant design changes will be made to accommodate the unique needs of the RAN, a firm fixed-price contract may not be flexible enough to accommodate the risk associated with these changes. In this case, a cost-reimbursement approach may be more appropriate to account for risk in the design and construction of the first-of-class, with a firm fixed-price contract for the follow-on vessels.

The types of contracts typically used for major shipbuilding programs in the United States are shown in Table 5.2. A cost reimbursement contract covers all of the contractor's allowable expenses up to a certain limit plus fee. This type of contract typically places a greater share of the risk on the government, which may end up paying more than the contract target cost. It is not unusual for cost-plus contracts to have high rates of unplanned cost growth. For example, in the LCS program the first two seaframes were awarded under a cost-reimbursement contract, and the program saw rates of cost growth

Table 5.2
Contract Types and Uses in Major Acquisition Projects

Contract Type	Description	Use in Acquisition
Firm fixed-price	Contractor is responsible for performance with a specified price	More appropriate for procuring off-the-shelf systems and platforms
Target cost incentive (fixed-price incentive firm target)	Contractor is responsible for performance under a target price with a target profit incentive.	More appropriate when more units are produced and there is an opportunity for the contractor to gain efficiency savings and the government to share in savings; more complicated to administer than a firm fixed-price contract
Cost-plus fixed fee	The contractor receives a predetermined fee	More appropriate for developmental programs or when costs cannot accurately be estimated; typically requires greater oversight and administration
Cost-plus incentive fee	The contract receives a smaller or larger fee based on how it performs based on cost or performance targets	More appropriate for developmental programs or when costs cannot accurately be estimated; typically requires greater oversight and administration
Cost-plus award fee	The contractor receives an award fee for meeting specific performance targets	More appropriate for developmental programs or when costs cannot accurately be estimated; typically requires greater oversight and administration

between 130 and 150 percent.[5] If Australia chooses to pursue a new design for the SEA 5000 program, it likely will have to pursue a cost-reimbursement contract, inasmuch as industry may be reluctant to accept the high levels of risk associated with a new ship design and build.

Appropriate use of contract incentives and awards fees can help to manage or mitigate unplanned cost growth in both firm fixed-price and cost-reimbursable contracts. The AWD contract uses a "pain-share gain-share" arrangement in which the incentive fees decrease toward zero as the direct project costs exceed the target cost estimate and increase when direct project costs fall below the target cost estimate.[6]

The U.S. Navy has had very good success in the DDG-51 program with a profit-related-to-offers strategy, to competitively allocate follow-on ship construction to two different shipyards. This strategy allows the shipbuilder that submits the lowest-cost bid for its "allocated ship" to receive a higher target profit percentage, and the shipbuilder that submits the lower bid for the next follow-on ship to be awarded an option to construct that ship as well.[7]

Allocate Contract Risk Appropriately Between Government and Industry

General defense acquisition wisdom recommends that contract type match program risks.[8] For example, the guidance documents for U.S. acquisition policy discuss the central issue of programmatic and technical risk and which party (government or industry) is better able to

---

[5] GAO, 2010.

[6] Australian National Audit Office, 2014, p. 117.

[7] GAO, *Arleigh Burke Destroyers; Additional Analysis and Oversight Required to Support the Navy's Future Surface Combatant Plans,* Washington D.C.: U.S. Government Printing Office, January 2012a, p. 23.

[8] See, for example, Office of the Deputy Director of Defense Procurement and Acquisition Policy for Cost, Pricing, and Finance, *Contract Pricing Reference Guide: Volume 4,* 2012; or Defense Acquisition University, *Defense Acquisition Guidebook,* January 10, 2012.

manage that risk as being a key determinant in the selection of type and incentives. Figure 5.1 summarizes the differences and broad preferences of contract incentives during the acquisition timeline.[9] The more uncertainty and risk favors cost-type contracts over fixed, variable, or target price. Earlier in acquisition also favors cost-type contracts over other forms, as well. Most important to note, there is no one best contract type for all circumstances. For an alliance approach, some form of risk-sharing contract is used as a way to align the parties, for example.

Best practices suggest that successful contracting strategies hold the contractor responsible for risks under its control (labor rates, productivity, material costs, etc.) and hold the government responsible for risks outside the contractor's control (inflation, requirements changes, legal changes, etc.). One important decision when establishing the acquisition strategy is deciding which equipment will be bought and managed by the government as GFE and which equipment will be

Figure 5.1
Spectrum of Contract Types and Cost Risk

Cost risk	High ◄——————————————————► Low				
Requirements definition	Vague ◄——————————————► Well defined				
Acquisition phase	Solutions analysis	Concept refinement	Design and build	Test and trials	Follow-on production
Contract type	Varied	CPFF, CPIF, or TCI	CPIF, TCI, FP	CPIF, TCI, FP	FP, TCI, VP

SOURCES: Adapted from Edward C. Martin, "Incentive Contracting," PowerPoint file, SAF/AQC Field Support Team, April 25, 2011, p. 9; and Defense Materiel Organisation, *Defence Procurement Policy Manual (DPPM): Mandatory Procurement Guidance for Defence and DMO Staff*, July 1, 2013.
NOTES: Contract definitions are per DMO, 2013 (U.S. terminology listed in parenthesis). FP = firm price (firm fixed price), VP = variable price (fixed price with economic price adjustment), TCI = target cost incentive (fixed price incentive fee), CPIF = cost plus incentive fee (cost plus incentive fee), CPFF = cost plus fixed fee (cost plus fixed fee).
RAND *RR767-5.1*

---

[9] From Martin, 2011, p. 9.

bought and managed by the contractor as CFE. Failure to appropriately assign responsibility at the contracting stage can result in later contract disputes over liability for increased program costs or delays. For example, the NAO found that in the Type 45 program, government responsibility for key equipment left it open to compensation claims upon delays.[10] Therefore, it is important to make informed decisions on which equipment is GFE rather than CFE. These decisions may be based on a number of factors, such as

- Who is better placed to manage the subcontractors (including technical and management capability or expertise)?
- At what point in the critical path of the program schedule is the equipment needed?
- Who is responsible for integrating the equipment into the system?

Construction schedules are tied to the planned delivery dates for key pieces of equipment. When making the GFE versus CFE decision, the government should consider its capacity to manage subcontractors and to ensure that equipment can be delivered within the schedule deadlines specified in the contract. Regardless of who is responsible for providing the equipment, it is important that the government closely monitor the health and performance of vendors to ensure that they remain certified and are delivering quality products.

Obtain Needed Intellectual Property and Technical Data Rights

A new ship program will likely stretch over several decades. Some of the ships acquired will still be operating more than 50 years from now. Given this long period of government commitment, it is important that the government have a plan for maintenance and support of these frigates. Successful support requires appropriate intellectual property (IP) rights to the technical data necessary to maintain these ships at

[10] NAO, 2009, p. 20.

best value to the government. Well-thought-out strategies for maintenance and IP rights are preconditions for successful interaction with industry.

Appropriate rights in technical data will have to be held by the government so that it can perform (or even allow competition for) maintenance on and support for the Future Frigates. With no rights to technical data, the government will be beholden to the original industrial suppliers for lifetime support of these ships. It is hard to conceive that this approach will represent best value.

The U.S. Defense Federal Acquisition Regulation Supplement has three categories of technical data rights:[11]

- Unlimited: The government can provide the data to anyone for whatever purpose. These rights are obtained when the government fully finances development of the technical data or when those data are in the public domain.
- Government Purpose Rights: The government can provide data to anyone else who needs them to perform work for or provide supplies to the government. These rights are obtained when technical data development funding is mixed and the government has a need for the data rights.
- Limited Rights: The government can provide the data to non-government parties only for emergency repairs, provided it limits access to just what is necessary for repair. These rights pertain when the technology has been developed entirely at contractor expense. The contractor must identify such data up front, mark the data with limited-right legends, and be able to defend the assertion of rights.

Notwithstanding the above, the U.S. government can negotiate, and if necessary pay, for greater rights in technical data if needed.

IP rights are particularly important when separate companies are contracted for design and for build and are not connected through a

---

[11] See U.S. Government, "Rights in Technical Data," *Defense Federal Acquisition Regulations Supplement,* Subpart 227.71, February 28, 2014.

formal alliance or information-sharing agreement. For example, the AWD program encountered IP challenges with its platform system designer, Navantia. The AWD design contract initially required Navantia to supply only two-dimensional (2-D) engineering drawings in PDF. These drawings proved difficult to interpret during production, and the alliance later decided that it would require the 3-D computer-aided-design (CAD) models. Navantia was unwilling to release the 3-D models because of IP reasons, requiring that the alliance negotiate the purchase of the 3-D models.[12]

A clean-sheet frigate design, funded entirely by the government, will give DMO the greatest leverage to obtain the technical data rights needed for operation and maintenance. In its prime design and build contract(s), the government will still need provisions that allow for the incorporation of components and systems from subtier equipment suppliers. It may need to acquire the appropriate data rights up front, find another supplier willing to grant greater rights, or pay for another organization to design a component or system with government funds and unlimited rights. Since the government will be operating these frigates for years, it is appropriate for it to be involved and require approval of these subtier decisions.

With a pure or evolved MOTS design, it is important that Australia have the IP rights to ensure that the ship can be properly modernized and maintained during its operational life. One problem that hindered the *Collins* program was the lack of the IP rights to the design of the basic platform and much of the fitted equipment. Not having the rights to *Collins* IP on future designs may constrain the design effort for the new submarine class that will replace the *Collins*. Although Kockums and the DoD reached a settlement in 2004 that provided ASC and its subcontractors access to Kockums' IP, the settlement still protected Kockums' proprietary rights. The McIntosh and Prescott report touched on the need for IP rights to be available to DoD, stating "(f)ailure to either own or have unfettered use of technology limits the

---

[12] Australian National Audit Office, 2014, paragraph 6.9.

alternatives open to the buyer when the supplier fails to produce and also more generally."[13]

A MOTS acquisition presents the greatest technical data rights challenge. By definition, Australia will not have paid to develop any of the technical data and will be in a weak position in negotiations to acquire the appropriate rights. This is why it is important to have an IP strategy up front, so that the cost of acquiring the appropriate technical data rights is factored in to the MOTS evaluation. If a MOTS selection could be made in the context of competition between MOTS alternatives, the likelihood of obtaining appropriate data rights at an acceptable price increases.

Even in the MOTS case, though, subtier equipment and systems could present a problem. A foreign shipbuilder could offer to grant unlimited rights to all the data it developed, but it could not provide Australia with rights to data it did not have in connection with equipment that suppliers had developed. The availability and cost of subtier equipment and system data rights will also have to be factored into any MOTS evaluation.

Evolved MOTS will obviously fall in between the two extremes discussed above with regard to IP challenges. But a well-thought-out IP strategy that supports a preconceived maintenance strategy will inform and facilitate dealings with industry and will ensure long-term support for the Future Frigates at best value to government.

Develop Processes for Managing Contract Changes

Despite best efforts in project planning before contracting, there is always some degree of uncertainty in large shipbuilding programs, and changes will invariably occur to the desired performance, technical specifications of systems or equipment, or organizational structures of industry partners responsible for designing, building, or testing the

---

[13] Malcolm K. McIntosh, and John B. Prescott, *Report to the Minister for Defence on the Collins Class Submarine and Related Matters,* Canberra: CanPrint Communications Pty Ltd., June 1999, p. 15.

platform. It is important to build into the contract processes for managing changes and to provide the appropriate management and legal structures to adjudicate changes as quickly and fairly as possible. These processes should have incentives for the timely adjudication of changes and metrics to track the status of each proposed change.

In all the shipbuilding programs we examined, there were significant cost overruns for the first ship of the class. This suggests that it also is important that programs reserve adequate contingency funds to address potential issues down the road. In the *Collins* program, for example, the contingency fund was approximately 2.5 percent. This low level of contingency funding hampered the ability of the program to deal with changes that arose later and contributed to strained relationships between the government and industry.[14] An adequate contingency pool for complex engineering projects is typically in the 10 to 15 percent range.

When funding is limited, it is especially important to impose program discipline on the number of engineering change orders and to consider the cost implications and trade-offs for the value gained from the change. The government can impose this discipline by maintaining an onsite presence at the design and build organizations. In that way, program managers can be made aware of issues that arise in the design and build and can take early steps to mitigate or adjudicate disputes.

Establish an Agreed upon Tracking Mechanism and Payment Schedule

Often cost and schedule problems arise that could have been mitigated if program managers had earlier indications of performance issues. Best practices in shipbuilding programs emphasize the importance of establishing tracking mechanisms to monitor cost and schedule performance and to tie payments to clearly defined milestones. It is important that the tracking system be based on metrics that are observable and measureable and that provide timely and accurate information to pro-

---

[14] Schank et al., 2011a, p. 18.

gram managers. The tracking system should also include independent validation mechanisms to confirm design and construction progress.

The *Astute* program had no effective tracking mechanism during its first years. This made it nearly impossible for the government or the prime contractor to recognize problems as they arose. The implementation of an earned value management (EVM) system by the MOD helped to bring the program back on track. EVM is a performance management system that has been mandated for use in the United States for most acquisition programs exceeding $315 million.[15] In U.S. shipbuilding programs, contractors are typically required to report EVM information at least once monthly.[16]

Best practices suggest that the RAN should require some core metrics to be reported by the shipbuilders to support EVM or alternative tracking processes:[17]

- actual cost of work performed
- budget cost of work performed
- budget cost of work scheduled
- estimate at completion
- budget at completion.

---

[15] EVM is a program management tool that "Integrates the technical, cost, and schedule parameters of a contract. During the planning phase, an integrated baseline is developed by time phasing budget resources for defined work. As work is performed and measured against the baseline, the corresponding budget value is 'earned.' From this earned value metric, cost and schedule variances can be determined and analyzed. From these basic variance measurements, the program manager (PM) can identify significant drivers, forecast future cost and schedule performance, and construct corrective action plans to get the program back on track. EVM therefore encompasses both performance measurement (i.e., what is the program status) and performance management (i.e., what we can do about it)" (U.S. DoD, *Earned Value Management Implementation Guide,* October 2006, p. 2).

[16] Contractor management systems must conform to the American National Standards Institute standards for EVM.

[17] Mark V. Arena, Hans Pung, Cynthia R. Cook, Jefferson P. Marquis, Jessie Riposo, and Gordon T. Lee, *The United Kingdom's Naval Shipbuilding Industrial Base: The Next Fifteen Years,* Santa Monica, Calif.: RAND Corporation MG-294-MOD, 2005b, p. 17.

Although EVM can help support schedule and milestone tracking, it is primarily a cost-tracking tool. As such, additional schedule control metrics should be developed and schedule updates and forecasts should also be reported by the shipyard.

Key Points with Acquisition and Contracting Strategies

Developing an appropriate contract structure for the level of program risk including carefully constructed contract incentives can help control costs and ensure that the relationship between the government and the various contractors avoids any disagreements or ill feelings. Some important lessons here include the following:

- *Include contract provisions to handle program risk that appropriately allocate risk between the government and industry.* Ensure that contractors are responsible for risks under their control and that the government is responsible for risks outside the contractor's control.
- *Consider IP rights and export rules.* IP sharing arrangements and military export rules should be clearly defined in the contract.
- *Develop processes for managing contract changes.* These should be specified in the contract, accounted for through contingency funding, and closely managed by the government through an onsite presence at the design and build organizations.
- *Establish an agreed upon tracking mechanism and payment schedule.* The mechanism should be supported by robust metrics and clearly defined milestones.

CHAPTER SIX

The Solutions Analysis Phase

This chapter describes the first step in any major acquisition program—the solutions analysis phase. It is during this phase that various solutions to the capability gap are identified and analyzed. This step is independent of the acquisition philosophy used. In fact, multiple acquisition options could be considered during this phase. And for each acquisition option, there could be multiple concepts (potential ship solutions). The objective of this phase is to narrow these concepts to a single concept to take forward. Any time during this phase, options or concepts could be eliminated as a result of performance or affordability reasons.

The nature of the issues and analysis does vary depending on the acquisition option, however. For pure MOTS or evolved MOTS concepts, this step must identify existing designs that could meet desired operational objectives and then evaluate the cost, schedule, and technical risks of those existing designs or modifications to them. In the case of an evolved MOTS concept, the identification of major changes to an existing design that are desired is a key activity of this phase. For a new design concept, basic conceptual engineering must be done to establish high-level characteristics of the ship. Regardless of the option, all candidate concepts for the next phase should have their general characteristics defined (measures of performance) along with a very basic conceptual-level design, weight statement, and life-cycle cost.

Set Operational Requirements

As noted in Chapter Four, the solutions analysis phase is the first formal program phase and perhaps the most important. Decisions made during this phase can have enormous cost and schedule implications later in the program. So it is critical to program success to understand the sensitivity of the cost and schedule of various design choices. The starting point for this phase is a set of defined and approved requirements from the operational community. However, these requirements are not set in stone. Trade-offs need to take place between requirements and cost so that an affordable solution can be achieved. This type of cost-effectiveness analysis is one of the key outputs of this phase (in a study called an analysis of alternatives [AoA]). The AoA study is used by decisionmakers to understand trade-offs and select a single solution that will be refined in the next phase. Another important deliverable is a technical development strategy to reduce risk on technologies that may be immature or are being applied in a military environment for the first time.

The operational requirements of the platform will be a key determinant in choosing the most cost-effective alternative for the program. Operational requirements include scenarios, concept of operations, and operational views (in the context of the broader defense force) for the operation of the ship. These requirements define the ship's missions and desired effectiveness in those missions. The operational requirements are later translated into performance requirements for such attributes as the ship's speed, endurance, reliability, and survivability during the early design. The operational requirements also set forth expectations for training, maintenance, and modernization connected with ILS. The platforms available under each acquisition option will have measures of performance that will contribute to how effective the platform is in meeting the various requirements.

Operational and performance requirements are often expressed in terms of *desired objective measures* (i.e., ones whose achievement is advantageous but not critical in all circumstances) and *minimum thresholds* (i.e., ones that are critical and that the platform must achieve without exception).[1] There are numerous requirements, and different platform options will measure up against them in different ways. Trade-offs between the operational requirements and a platform's cost, schedule, and risk are typically needed when evaluating various platform options. These trade-offs dictate that the various requirements be prioritized. One additional lesson from weapon system development in the United States is that the requirements should not be over-specified at the beginning of the design activities such that there is no ability to make trade-offs between ship attributes and cost.

A set of well-described operational requirements is the *entry point* for the acquisition process. Understanding the requirements will involve some back-and-forth between the operations and technical communities. An important aspect to this back-and-forth is for the technical community to provide feedback to the requirements community concerning the affordability of their choices. Also, requirements should not define the system characteristics (e.g., how large a hull or how fast it goes); rather, they should define the operational needs of the ship (e.g., the ability to locate and interdict enemy vessels, the ability to rescue downed pilots, nominal crew size). Programs with poorly defined requirements often experience schedule slip and large cost growth.[2]

Along with the requirements, the operational community must define how the system will be employed and supported. In the United States, the CONOPS document is a formal statement of the operational vision. The document describes how the system will be used for actual missions (along with supporting scenarios or vignettes of operation) as well as normal day-to-day operations. It also describes deployment patterns and locations, basing, and maintenance assumptions to be used during concept exploration and refinement. The CONOPS is not a formal statement of requirements or a ship specification. It is meant to guide the acquisition community in defining needed performance characteristics and features. Along with requirements documentation,

---

[1] Threshold and objective capability requirements are used in the United States for ship acquisitions but are not commonly used in the Australian capability development process.

[2] Bolten et al., 2008.

it is one of the earliest program documents. In Australia, the requirements development process begins within the navy. The navy initially develops a capability needs statement, which is subsequently formalized by the capability development group in the operational concept document.

For naval ships, selection of requirements happens in programs' earliest phases. These operational requirements are translated into performance specifications that have clear implications for technology development and technical risk. Because unrealistic technical achievability requirements can lead to later cost growth and program delays, capability needs must be weighed against affordability goals. Operational requirements, particularly the desired operational availability, also have implications for integrated logistic support planning and the total life-cycle costs of the program. Readers interested in requirements milestones should refer to the discussion in Chapter Four.

Identify Important Considerations

Analysis of Alternatives

An AoA is a systematic, independent, and unbiased analysis of a prenarrowed set of alternative ways a mission need might be satisfied. It should assess the alternative materiel solutions, including associated technology maturity and technical risk. AoAs should make a case for the most cost-effective alternative(s) and the capabilities and utility that acquiring the most cost-effective alternative(s) will provide. However, a single best solution may not emerge. In such cases, an AoA should demonstrate the various trade-offs between alternatives so that decisionmakers can prioritize.

Typically, the first phase of an AoA involves the development of a study plan for the analysis. This plan states the ground rules and assumptions of the study, including the required capabilities, operating environments and threats, broad categories or alternatives, metrics for evaluation, measures of effectiveness, and cost analysis methodologies. It serves as the roadmap for an AoA and defines how the goals will be met and when findings will be provided.

Industrial Base Resource Assessment

In this early phase, it will be necessary for the SEA 5000 program to assess what industrial capabilities exist (both design and construction) and begin to make plans for the program on how it will leverage domestic and international industry. Part of the assessment needed is to evaluate not only what capabilities exist but how available they will be to the SEA 5000 program. For example, other shipbuilding programs such as SEA 1000 might compete for resources. Is there enough capability for both programs? Would adjustments to the schedules for both programs de-conflict resource demands? It may be necessary for the DMO and RAN to set priorities across the entire build program portfolio and avoid suboptimizing at a program level. The UK had a similar issue during the mid-2000s. A series of conflicting shipbuilding demands caused several programs to shift their planned starts. And the lack of a coherent plan made it difficult for industry to right-size.[3]

Future Upgradability and Margins

One important early decision is how to accommodate and plan for change over the life of the class. Surface combatants can last upward of 30 years; so to remain relevant, ships must be affordably upgraded and improved several times over their lifespan. Some changes might be easy to accommodate, such as a software upgrade. But other changes might require more extensive modifications to the ship, such as adding an additional set of equipment. Additional space, power, and cooling (heating, ventilation and air conditioning [HVAC], or water) must be available to make the installation of new equipment functional. Moreover, weight margin (in terms of total weight and stability) must also be present to accommodate the new equipment and foundations. To simplify, we can think of two different approaches to making the ship available for future upgrades. One approach is standard interfaces and is typically used with C4I components. For HM&E systems, adaptability is achieved through margins (beyond what is needed at ship delivery).

---

[3] Arena et al., 2005b.

For C4I, future adaptability is achieved through standard interfaces and spaces. Such interfaces span a broad range of implementations from such simple specifications as hookups for power and water to more sophisticated approaches in terms of standard equipment racks with specific space and utilities hookups to reconfigurable spaces or rooms in ships where equipment can be flexibly hosted. For example, the newer U.S. aircraft carriers have reconfigurable spaces to adapt to the changing needs of the Navy.[4] Ventilation and wire ways run underneath a false deck so that equipment can be easily connected and disconnected. An overhead grid system creates compartmentalization and stabilization points for equipment racks and also for such mount items as lighting, monitors, and speakers. The deck and overheads have a rail system that will allow the equipment to be bolted down instead of being welded down. HVAC systems are located in adjacent spaces. This additional flexibility may require some additional up-front cost and must be a conscious early decision, but there might be considerable through-life savings.

For HM&E, future adaptability comes through including additional capacity at delivery (such as more power generation than needed initially). This extra capacity is termed *margin*. It is typically very difficult and expensive to change HM&E once the ship is designed and built. So, it is far more cost-effective to add additional capacity from the start of design rather than to plan for an exact amount, only to have to extensively redesign the ship later. The HM&E margin is set differently corresponding to acquisition (builder's) and service life allowances (SLAs). The acquisition margin is to compensate for uncertainties during the design, procurement, and build processes and is focused on weight. For example, the exact weight may not be known for a particular piece of vendor equipment. If the equipment is heavier or takes more power than initially planned, some of the margin can be used for this growth. The weight margin is important to ensure that the ship will be able to perform to its specifications throughout its life. Addi-

[4] William A. Deaton and James L. Conklin, "Developing Reconfigurable Command Spaces for the Ford Class Aircraft Carriers," *Engineering and Total Ship Symposium 2010*, American Society of Naval Engineers, June 2010.

tional weight increases the ship's draft, which adversely affects piloting, speed, and range.

Another important margin related to weight, and its placement on the ship, is stability—the ability of a ship to operate safely at sea and recover from roll and pitch conditions. The key measure of stability is metacentric height (GM). This is the height between the metacenter and the center of gravity. The metacenter is a naval architecture concept relating to the changing locations of the center of buoyancy as the ship rolls and pitches. The greater the GM, the more a ship is stable in a seaway. The greater the height of the center of gravity above the keel (KG), the less the GM. Thus, increasing a ship's KG decreases its stability.

Typical weight margins used by the U.S. Navy are listed in Table 6.1. Table 6.2 displays the corresponding KG margins. These are presented in light ship displacement (full-load displacement less the weight of crew, stores, fuel, and ammunition) as this represents the weight the shipbuilder contributes to the ship.

The wide range of margins (as represented by their standard deviations) reflects, in part, sensitivity to the design risk. Follow-on designs

Table 6.1
Notional Acquisition Weight Margins (As a Percentage of Light Ship Displacement)

Margin Account	Mean	Standard Deviation
Preliminary and contract design	0.8	4.4
Detailed design and build	4.5	9.8
Contract modification (contract changes)	0.4	2.1
GFE/GFM	0.2	0.7
Total	6.0	17.5

SOURCE: Society of Allied Weight Engineers, Inc., *Weight Estimating and Margin Manual for Marine Vehicles,* Marine Systems Government—Industry Workshop, Society of Allied Weight Engineers, Recommended Practice Number 14, May 22, 2001.

NOTE: Total margin is not a simple summation of the individual phase averages.

Table 6.2
Notional Acquisition KG Margins (As a Percentage of KG in Light Ship Displacement)

Margin Account	Mean	Standard Deviation
Preliminary and contract design	2.7	6.1
Detailed design and build	1.7	5.1
Contract modification (contract changes)	0.3	1.9
GFE/M	0.1	0.4
Total	4.8	14.5

SOURCE: Society of Allied Weight Engineers, Inc., 2001.
NOTE: Total margin is not a simple summation of the individual phase averages.

with minimal changes might have very low margins whereas new (clean-sheet) designs will have much higher margins applied. Translating to the SEA 5000 program circumstances, a pure MOTS solution will have no acquisition margin because the design exists, whereas a clean-sheet design might have much higher margins than the means listed in Tables 6.1 and 6.2. The amount of margin for a modified MOTS approach is unclear as it depends on how much redesign takes place.

SLAs depend on the type of vessel and projected service life. The more frequent the upgrades and longer service life result in greater allowances. Table 6.3 shows notional values for SLAs for various ships used by the U.S. Navy, based on historic data.

Allowances for other systems, such as utilities, are less well defined. The U.S. Navy's rule-of-thumb is that electric load should grow 1 percent per year for the first two-thirds of a ship's life cycle and should not grow during the remaining one-third of its life.[5] For the SEA 5000 program, the appropriate SLA will be less about the option and more

[5] Jonathan Page, *Flexibility in Early Stage Design of US Navy Ships: An Analysis of Options*, master's thesis, Engineering Systems Division and the Department of Mechanical Engineering, Massachusetts Institute of Technology, Cambridge, June 2011.

Table 6.3
Service Life Allowances for Weight and KG at Delivery

Ship Type	Weight (percentage)	KG (meters)
Combatants	10.0	0.30
Carriers	7.5	0.76
Amphibious		
Large deck	7.5	0.76
Other	5.0	0.30
Auxiliary	5.0	0.15
Special ships and craft	5.0	0.15

SOURCE: Society of Allied Weight Engineers, Inc., 2001.

NOTES: Weight percentages are based on the predicted full-load displacements at delivery. KG values are based on predicted full-load departure KG at delivery.

about the future upgrade plans. However, a pure MOTS solution will have the SLA mainly fixed and the RAN will have to live with whatever margin exists for the design.

Computing Architectures

A related issue to future adaptability is the choice of computing architectures for the ship. The traditional federated approach decentralizes the computing hardware and software functions of the various C4I systems while allowing them to share data and information through a common network. In this way, hardware or software problems with one C4I system, or an upgrade to a system, can be isolated and addressed without affecting the performance of other C4I systems. Another advantage of this approach is widening the vendor base that provides the C4I/computing capabilities. Various systems can be procured from different providers and "plugged" into the federated system through a common set of specifications.

One downside of using a federated approach is that it includes some redundancies in computing hardware and software that raise costs and place increased demands on ship services. Another disadvantage is the need to establish and maintain a set of rigid specifications that govern the connection of the C4I systems into the federated architecture.

An alternative approach using integrated architecture reduces hardware and software redundancies by using both a shared network and shared computing hardware and system software. Computing functions in integrated systems are typically software programs that use the common system hardware processing and the common network for sharing data and information. However, integrated systems can potentially lose all C4I functions when there is a problem with the common hardware or system software; there also may be fewer suppliers of the more complex hardware and software systems available. Additionally, the government must maintain and control the computing interface standards and libraries. And it must actively maintain the "middleware" software that translates and passes information into a common format.

Setting requirements is an important step in the acquisition process and choices here can have large downstream implications. Lessons from previous programs suggest that operational requirements should be realistic, clear, stable, and testable. We next discuss some of the best practices and lessons learned for program managers in developing and managing operational requirements for shipbuilding programs.

Involve All Appropriate Organizations in Setting Operational Requirements

The program office must be supported by adequate technical, operational, and management expertise. This is especially important when setting requirements early in the program. Technical experts in laboratories and test centers can keep the program manager informed about existing and new technologies. The RAN can provide insights into current naval force missions and capabilities, and the organizations that maintain ships can provide information on how designs and operational requirements influence support costs. Experienced designers and builders can shed light on the difficulties and costs of achieving certain operational objectives. Moreover, these experienced designers and builders can help program office engineers and acquisition experts draft contract specifications that balance these demands. Additionally, they can offer advice to achieve a coherent requirement set that specifies desired performance and safety outcomes in a manner clearly understood by all parties.

The important issue is that the program manager and other decisionmakers understand the trade-offs between the cost, performance, and risks of technical choices when setting requirements. The technical organizations and the operators, builders, and maintainers must be able to effectively show the implications of different operational requirements to allow the program manager to make design and operational trade-offs in a structured and coherent manner.

Involving various organizations is important throughout the life of the program. The program manager should have decisionmaking authority and must draw on various technical and operational resources to make those decisions. Also, involving all appropriate organizations helps develop program managers who are knowledgeable and experienced.

Remember That the Ship Is an Integration of Various Systems

A ship is an integration of the hull, a power and propulsion system, sensor and communication suites, and weapon systems. Operational requirements in one area will affect design considerations in the other areas. More capable sensor systems may require additional power and a different propulsion system, which could affect the basic hull design. The desire for greater weapon capability with more or newer weapons may also affect hull dimensions.

It is challenging to find the right balance among the various system requirements, especially when doing so for a ship class that will be in the operational fleet for 30 years or more. Operational requirements and technologies change over time resulting in major modifications during a ship's life. When setting requirements for different systems, program personnel must understand current and emerging technologies in those systems, how requirements might change in the future, and trade-offs between costs and risks.

It is important for program managers not only to know the current state of various technologies but also to understand how changes to operational requirements relate to the technologies that are available. That is, if certain operational goals are beyond the state of current technology, what operations can existing technologies support? This relates to trade-offs between operational requirements and technological risks (and costs). Again, this area is where both operators and the technical community are important during the early stages of a program. The program must understand technical boundaries and the risks inherent in an evolutionary versus a revolutionary strategy. Existing systems can be scaled to some degree. However, scaling an existing system too far leads to difficulties and ultimately results in entirely new systems or significant problems. Also, integrating existing systems may be more challenging than anticipated. For *Astute*, the MOD and the prime contractor greatly underestimated the effort involved in integrating various systems and equipment from previous classes of submarines.

Even when the operational requirements for a new class are similar to those of the previous class, program managers need to be kept informed of the continued ability to deliver the level of technology needed. With the long operational life of modern ships, equipment and system obsolescence is a major driver of change and risk. Obsolescence risk is often compounded by change in safety or legislation that makes legacy systems and equipment noncompliant.

Clearly State Requirements

Operational requirements must be clearly stated and be an appropriate mix of key performance requirements and technical standards. A myriad of requirements, specifications, and standards can at times be conflicting and difficult to interpret. The operational requirements must be clearly stated as the desired performance of the ship in various key areas. Key areas include speed, payload, signatures, and such other key characteristics as crew complement and operational availability. These performance requirements must be backed with some level of technical specifications, especially in the area of safety. Requirements specification is a difficult balance of staying within known and approved standards and allowing design innovations (especially innovations that aim to reduce costs). The operational requirements should be supported by standards that relate to different functional systems. The program should allow the prime contractor to challenge standards and specifications if it can prove that the change will reduce cost or improve performance at equal or lesser risk. The RAN must have knowledgeable and experienced personnel to objectively evaluate the contractor's change proposals.

Develop a Test Plan for Requirements

Program managers must understand that in addition to specifying an operational requirement they must also spell out how to test for the achievement of that requirement. The "hands-off" approach taken by the MOD during the initial stages of the *Astute* program led to (or resulted from) the deactivation or downsizing of the Royal Navy and MOD technical organizations that had overseen the testing and commissioning of all prior UK nuclear submarines. Without this knowledge and expertise, testing was largely ignored during the contract negotiations and early stages of the program. Planning for testing and commissioning did not begin until approximately five years after the contract was signed. With the first of class, both parties struggled to identify and approve procedures to test whether the vessel met performance requirements and to completely understand the time required to test the new capabilities and design.

Stating an operational requirement is the first step in setting program goals. But that first step must be complemented by a plan to understand whether the platform meets the requirement. This typically involves test procedures—who will test, how the test will be conducted, and how success or failure will be measured. Although it is often difficult to plan tests early in a program, doing so ensures that all parties agree on the processes to measure how the platform meets operational capability objectives. Incremental testing of equipment before it becomes part of a system and before that system is inserted into the hull should be encouraged.

Keep Requirements Stable

Ideally, a shipbuilding program would enter the design phase with stable operational requirements. However, given the long lead time on ship design and construction, there is a danger that requirements may change as a result of changing threats or operational circumstances. This "requirements creep" has historically led to contract disputes, cost overruns, and schedule delays across large acquisition programs. A 2008 RAND study of cost growth in major defense acquisition programs found that approximately 13 percent of total program cost growth can be attributed to government changes to requirements, with most of the cost growth occurring in the development phase.[6]

When requirements change as a result of legitimate capability needs and are fully supported through existing contract contingencies or additional funding, this may not create problems for the program. However, requirement changes that are not validated, or that are made after a majority of the detailed designs are complete, may not provide an efficient trade-off of cost for capability. For example, in the AWD program, design changes that were made after the First Pass resulted in an estimated additional $122 million to the total program cost.[7] When requirements are changed, it is important that the program office understand the full cost and schedule implications of those changes. It is important to set the requirements early and avoid changing them unless there is a clear and compelling need. Keeping requirements stable depends on certain management decisions and consider-

---

[6] These were changes made by the program office, lead service, Department of Defense, or the legislative bodies, see Bolten et al. 2008, p. 27.

[7] Australian National Audit Office, 2014, p. 35.

ations in the earliest phases of concept development as well as program discipline throughout the design and construction phases.

One strategy for mitigating requirements creep is to conduct batch builds where technology is allowed to mature between batches. This would allow a minimum capability to be achieved for the first batch of ships in the class, whereas affordability and capability trade-offs would be considered in follow-on batches. Modular and open architecture design solutions can also help to support changing combat system requirements in future batch buys.

Support Requirements Through Relevant Technical, Safety, and Classification Standards

A final consideration in developing operational requirements involves gaining an understanding of how they will be met by the requisite technical, safety, and classification standards.[8] Classification standards are established by independent classification societies to establish and maintain technical rules and regulations for the design and construction of maritime vessels. As these standards are typically focused on commercial shipping, it is important that any classification standards adopted are consistent with military needs for survivability and safety. The *Queen Elizabeth* carrier program is the first class of UK warships that has been designed from the start using commercial shipbuilding standards, namely, the Lloyds Naval Ship Rules for Systems and Structural Design. There have been some questions as to whether the commercial-grade materials and components (for piping, brackets, etc.) specified in these standards are acceptable for military application.

It also is important that program personnel understand that the standards are stable, mature, and consistent before initiating any design work. The LCS program intended to implement a new set of construction standards set out by the American Bureau of Shipping called the

---

[8] John F. Schank, Frank W. LaCroix, Robert E. Murphy, Mark V. Arena, and Gordon T. Lee, *Learning from Experience, Volume I: Lessons from the Submarine Programs of the United States, United Kingdom, and Australia*, Santa Monica, Calif.: RAND Corporation, MG-1128/1-NAVY, 2011d.

Naval Vessel Rules (NVR).[9] However, the full set of approved standards was not published until after the design and construction contracts for the first two ships had been awarded.[10] Furthermore, the Navy modified the NVR after contract award to increase the warship's survivability. These changes during design and construction have created a number of issues for the program and have been cited as one reason why the LCS costs have grown dramatically.[11] What is not clear is whether this modification to NVR was a result of a requirements change or the fact that NVR was not fully mature and understood by the Navy.

Key Points for the Solutions Phase

Decisions made early regarding the desired operational performance of a new ship influence the technology risk for the program and the likelihood of its success. Operational requirements for the platform are translated into performance specifications that lead to technology choices to achieve the desired performance. Those operational requirements, especially the desired operational availability, also affect integrated logistics support planning. Important lessons here include the following:

- *Involve all appropriate organizations when setting operational requirements.* By engaging a range of experts in the development of operational requirements, program managers can ensure that requirements are technically feasible, testable, and sustainable throughout a ship's life cycle.

---

[9] These standards were intended to be a hybrid between commercial and military standards.

[10] Delores Etter, Paul E. Sullivan, Charles S. Hamilton, and Barry J. McCullough, *Statement before Subcommittee on Seapower and Expeditionary Forces of the House Armed Services Committee on Acquisition Oversight of the U.S. Navy's Littoral Combat Ship Program*, February 8, 2007.

[11] Ronald O'Rourke, *Navy Littoral Combat Ship (LCS) Program: Background and Issues for Congress*, Congressional Research Service, RL33741, August 4, 2014a, p. 38

- *Remember that the ship is an integration of various systems.* When setting requirements for different systems, program personnel must understand the current and emerging technologies in those systems, how those technologies will be integrated, how requirements might change in the future, and the trade-offs between costs and risks.
- *Clearly state operational requirements as a mix of key performance requirements and technical standards.* Have the discipline to avoid changing requirements unless there is a clear need for the change, and ensure that there is a sound understanding of the effect on cost and schedule of requirements changes.
- *Understand that operational requirements also must specify how to test for the achievement of that requirement.* Although it is often difficult to plan tests early in a program, it is necessary to ensure that all parties agree on processes to measure how the performance of the platform meets operational capability objectives. Incremental testing of equipment before it becomes part of a system and before that system is inserted into the hull should be encouraged.
- *Keep requirements stable.* Avoiding requirements creep requires early planning to ensure that processes are in place to avoid unnecessary change orders and to manage the cost and schedule implications of necessary changes.
- *Operational requirements should be supported by relevant technical, safety, and classification standards.* The standards should be consistent with military needs and be stable, mature, and consistent before design and construction.

CHAPTER SEVEN

Design Activities

This chapter describes various design activities that a project goes through, beginning with the concept design and ending with the detailed design that provides the product model, construction drawings, and procurement specifications for material, equipment, and systems. The chapter also describes various considerations that must be addressed during each step in the design process.

Concept Refinement/Design

Definition and Activities

During this initial step in the design process, mission needs are defined, desired platform operational characteristics are explored, future threats are examined, research and development efforts are proposed, and basic cost and schedule estimates are established. Concept design point studies are initiated to analyze and compare the effects of different platform capabilities with ship characteristics and cost.

The objectives of the concept design phase are threefold. First, the concept phase forms the basis to begin restricting the performance and operational characteristics of the platform, thereby codifying requirements invoked on the designers. Second, performance gaps are identified in current technologies that require research and development efforts to mitigate. Finally, a basic cost and schedule assessment is completed during concept design.

A relatively small cadre of naval architect designers, cost engineers, selected technology subject-matter specialists, and ship operators generally accomplishes concept design point studies. The actual number and skills needed are determined by the type complexity of the ship and underlying technologies.

Important Considerations

Maintenance and Basing Strategy

The CONOPS document (produced early in the previous phase), defines a draft plan to base and maintain the ship. During concept refinement, more detailed plans on maintenance and modernization planning should be developed—such as a notional docking and overhaul plan. The comparative amounts and types of maintenance performed by the crew and by supporting contractors ashore are defined during this phase. A logistics support plan should be developed along with a spares and replacement parts strategy. Also, responsibilities for logistics and support activities should be defined between government, DMO, RAN, and industry. The frequency and periods of modernization should also be established for each major system. Last, needed infrastructure to support the operation of the class should be defined and costs estimated.

Crewing Strategy

Modern navies are aggressively reducing crew sizes on new ships as a way to reduce ownership costs. Personnel costs tend to be one of the largest components of life-cycle cost. As a recent example, the U.S. Coast Guard is attempting to reduce crews sizes to a level that is roughly two thirds that of the levels for legacy ships of similar size. Maximum crew size is difficult to precisely define at the conceptual stage as it can be driven either by typical workload (watch standing, maintenance, training, etc.) or by when the ship is under battle conditions (including the need for damage-control parties and so forth). Being overly aggressive in reducing the crew size has the potential to cause shortfalls in operational effectiveness or materiel readiness. When the crew is overworked, important activities may not be accomplished (such as basic maintenance or boarding operations). This shortfall could also lead to safety issues if the crew becomes fatigued.

The U.S. Navy's recent experience with the LCS is a good example of the challenges with an aggressive crew plan. For trial operations, 20 additional crewmembers above the standard complement of 75 individuals were needed for counterdrug operations.[1] Although smaller crew complements can lead to lower operational costs (through reduced numbers of personnel), they can cause challenges in other areas such as maintenance and training. It is relatively inexpensive to add a few additional berths and storage space to a ship during the conceptual stage of design. Additional rack space hedges against a potential shortfall in crew numbers and allows future flexibility in carrying additional detachments, if required.

Design Standards/Classification Societies

According to the International Association of Classification Societies,[2]

> The purpose of a Classification Society is to provide classification and statutory services and assistance to the maritime industry and regulatory bodies as regards maritime safety and pollution prevention, based on the accumulation of maritime knowledge and technology.
>
> The objective of ship classification is to verify the structural strength and integrity of essential parts of the ship's hull and its appendages, and the reliability and function of the propulsion and steering systems, power generation and those other features and auxiliary systems which have been built into the ship in order to maintain essential services on board. Classification Societies aim to achieve this objective through the development and application of their own Rules and by verifying compliance with international and/or national statutory regulations on behalf of flag Administrations.
>
> The vast majority of commercial ships are built to and surveyed for compliance with the standards laid down by Classification

---

[1] See, for example, Ewing, 2009, or GAO, 2010.

[2] See International Association of Classification Societies, *Classification Societies—What Why and How?* June 2011.

Societies. These standards are issued by the Society as published Rules.

A choice for the SEA 5000 program will be whether to classify the Future Frigate, and if so, under which classification society. This choice will be constrained by the acquisition option selected. With a pure MOTS or evolved MOTS, the ship will likely have been designed (and built) using a particular society's standard. Changing to a different standard could involve costly reanalysis and potential redesign. So it may be more cost-effective to stick with the original classification society. For a new design, the choice of society is open.

Need for New Technologies

An important early trade-off that takes place is the amount and level of new systems and technology that will be incorporated into the design. New technologies can provide an operational edge and help to maintain the design's relevance to future threats longer into the future. Some new technologies allow for lower operation and sustainment costs either through more reliability or smaller crew complements. Regardless of the reason for choosing new technologies to incorporate, such technologies come with greater cost and schedule risk than more mature or proven technologies. Thus, the RAN will have to carefully balance future needs against risk in making choices. As we discuss in the following pages, new technologies generally require prototyping or testing during the preliminary design phase to reduce risk, which necessitates additional, up-front cost.

The ability to incorporate new technologies will be constrained, to some degree, by the design choice. For a MOTS solution, there will be limits to what technologies can be incorporated. With a clean-sheet design, there are far fewer constraints. Any new technologies will need to be prototyped and tested during the next phase, and this activity may include building a testing facility.

Preliminary Design

Definition and Activities

In this stage, the fidelity of the concept design is improved; individual systems are notionally sized and arranged within the hull. Performance characteristics are validated, and technologies requiring research and development are initiated. The traditional engineering and science disciplines—encompassing structures, acoustics, systems, shock, and hydrodynamics—are fully involved in the process to confirm the acceptability of the design in meeting requirements.

Important Considerations

Prototyping and Testing

During this phase, activities begin to prototype new or unproven technologies that will be used on the ship. For example, the prototyping might be as simple as testing a new piece of equipment in a maritime environment (on an existing hull). At the other extreme, prototyping and testing might involve building an integration facility to test complete major systems; this is more often done for C4I, weapons, and propulsion systems. Such facilities not only serve to test individual components but also help to validate that the entire system works as designed.

It is also at this stage that design teams are formed. Where a limited number of government and RAN personnel, augmented by private-sector subject-matter specialists, are involved with the conceptual design, preliminary design sees growth in designer and engineer numbers, typically from private-sector firms. For a new design, the design team can grow to several hundred people. For a pure MOTS option, the preliminary and contract designs are basically finished but there is still the need for government and RAN designers, engineers, and managers to understand the design and the systems that will be on the ship. An evolved MOTS option will involve some private-sector and government/RAN people; the number can approach the levels for a new design if the changes to the base design are significant.

Essential and Important Performance Requirements

The identification and quantification of key[3] performance requirements[4] is one of the important products of this design phase (although some effort may spill into the next phase—particularly for the desirable requirements). "Essential" requirements are the performance attributes of the system that are integral in meeting the mission requirements. Typically for shipbuilding programs, essential requirements will be part of the contract solicitations. These requirements must be measurable and quantifiable (e.g., how fast and how far) and not qualitative. "Important" requirements are those that are viewed as not being mandatory to meeting the mission requirements but still desirable. These important requirements also may appear in ship contract solicitations to inform industry but will not have the same contractual weight.

Two levels are specified for each key requirement: threshold and objective. Threshold values are the minimum acceptable value achievable at low-to-moderate risk, and an objective value fully meets the desired operational goal but at higher risk in cost, schedule, and performance.[5] Only a few essential and important requirements should be selected. Otherwise, the system might become unaffordable or too technically challenging to design.

Contract Design

Definition and Activities

The most important output from this phase is concurrence and agreement in interpretation of platform capabilities, demonstration of a level of fidelity in design maturity, and confidence in the capability of research and development tasking to deliver performance such that detailed ship specifications can be written. The design of the platform

---

[3] Those deemed "essential" and "important."

[4] These are analogous to U.S. KPPs and KSAs.

[5] Adapted from Australian Department of Defence, Defence Materiel Organisation, *Defence Materiel Handbook (Engineering Management): Defence Capabilities Document Guide,* Version 1.0, DMH (ENG) 12-3-003, November 2011.

is completed, and technological risks reduced, to the extent that a fully priced contract can be awarded, competitively or otherwise, for detailed design or follow-on construction. The specifications may be detailed and directive in nature and specify how to accomplish the requirement, or they may be performance-based and describe what capabilities are desired but not specify how to engineer systems to obtain them.

Important Considerations

Designer/Builder Relationships

It is important that the ship designer and the ship builder stay closely tied during the design process. As described in Chapter Four, in the concurrent design-build model, the builder is an integral part of the design team, providing insights into the construction issues that surround specific aspects of the design. This close relationship is easily attained when the designers and builders are within the same organization. It becomes more difficult but even more important when the shipbuilder is not part of the same organization as the ship designers. The ship designers must understand the capabilities of the shipyard workers and facilities and the processes they use during ship construction. And, the shipbuilders must understand the specifics of the ship design and how their facilities and processes may need to be adjusted to execute the designers' plans.

Testing and Commissioning Plan

Besides great technical and design definition, planning must begin during this phase for the testing, trials, and commissioning of the ship. Most surface combatants have complex weapon and mission systems that must be proven before delivery. A sequential testing process (described below) is a strategy to complete these activities successfully. Often, additional resources will be needed during testing and trials (e.g., additional facilities, test weapons, and targets). Funding for these supporting resources must be planned and budgeted.

Detailed Design

Definition and Activities

This final design phase produces all the documents, drawings, test procedures, and schedule relationships to allow the construction, outfit, and test of the platform. Typically, construction starts before all drawings are complete. If this gap between start of construction and the completion of drawings is too great, problems can occur. Construction is limited by the drawings that are available, and changes to arrangements and specifications can lead to costly rework during the construction process.

Important Considerations

Continuing Support to Construction

The requirement for design resources does not end with the completion of the detailed design phase. Designers and engineers at the design contractor continue to support the construction of the platform through the total build of the class. Modifications to the initial designs are often needed to correct errors, address new missions and new equipment, or support manufacturing process changes to reduce the cost of building the platform. Government personnel also are needed during construction to work with the design organization or builder on design changes and to approve any changes. Government personnel also must monitor the construction of the platform to ensure that the delivered ship meets all requirements and can operate safely.

Building the Ship in Batches

Often, when there are a number of ships in a new class, the ships are built in different batches (also called blocks or flights). For example, the U.S. DDG-51 class of destroyers, first commissioned in 1991, has been built in three flights with a fourth flight now being designed. Also, the *Virginia*-class submarines are built in blocks with each block having some upgrades to the ones in the previous block. Building ships in blocks allows new technologies or systems to be integrated into the basic HM&E design. Given that the Future Frigate may be in construction for 16 or more years, there likely will be a need to upgrade the initial design somewhere around the middle of the ship's operational life.

It is important during the design phase to recognize that technology and mission changes during the life of a ship class may necessitate the need to build different flights or to update and upgrade systems on ships in earlier flights that are in service. The design phases must aim for adaptable ships or ones that can be modernized quickly and at low cost. Adaptability in ship designs typically mean flexibility and modularity as well as providing easy access to systems and equipment that have a high probability of being replaced during the 30 or more years the ship will be in service. Flexibility in a ship design often implies more space and higher margins for power, cooling, and bandwidth.[6]

Required Resources

Personnel—Skills

A wide variety of skills and professional technical expertise are used during the design of naval ships. In prior work for the U.K. MoD and U.S. Navy, RAND identified three broad categories of technical skills: (1) designer (meaning the technical workforce that generates the design), (2) professional engineers (encompassing people who are responsible for technical analysis and validating that the design is safe and can meet requirements), and (3) technical managers (constituting individuals who are responsible for oversight of the design process).[7] In Table 7.1, we list and define the various technical skills involved in each category.

---

[6] See John F. Schank, Scott Savitz, Ken Munson, Brian Perkinson, James McGee, and Jerry Sollinger, *Designing Adaptable Ships: Modularity and Flexibility in Future Ship Designs,* Santa Monica, Calif.: RAND Corporation, RR-696-NAVY, forthcoming.

[7] See Hans Pung, Laurence Smallman, Mark V. Arena, James G. Kallimani, Gordon T. Lee, Samir Puri, and John F. Schank, *Sustaining Key Skills in the UK Naval Industry,* Santa Monica, Calif.: RAND Corporation, MG-725-MOD, 2008.

Table 7.1
Technical Skills in Warship Design

Skill Category	Major Skill Area	Detailed Subskills
Designer	Electrical and control	Electrical system component, electrical analysis, electrical design, power generation
	Mechanical/fluids	Mechanical component, mechanical system, mechanical design, piping design, HVAC, fluid system design, hydraulic system design
	Hull/structural/ arrangements	Structural engineering, structural arrangement, structural design
	Other detailed design	Engineering support, life-cycle support, software engineering, information technology (IT) support
Professional engineers	Acoustics/signatures/ dynamics	Signature analysis, shock analysis
	Combat systems and integration	Combat system integration, combat system design
	Electrical and control	Electrical system component, electrical analysis, electrical design, power generation
	Mechanical/fluids	Mechanical component, mechanical system, mechanical design, piping design, HVAC design, fluid system design, hydraulic system design
	Naval/marine architecture	Naval architect, marine engineer, weights analysis, standards
	Hull/structural/ arrangements	Structural engineering, structural arrangement, structural design
	Testing, commissioning, and acceptance	
	Safety and environmental	Safety engineers, environmental engineers
	Welding/metallurgy/ materials	
	Propulsion	Shafting and gear design, prime mover analysis, propeller design and analysis
	Other engineering	Engineering support, life-cycle support, software engineering, IT support
Technical managers	Planning and production support	Scheduling, purchasing support, component support
	Program management	Program management, schedule and cost control, estimating

SOURCE: Pung et al., 2008.

Product Models[8]

The complexity of the ship design process places a great demand on information technology systems to support design, in particular systems that enable CAD/CAM. The most complex of these tools is the 3-D product model, which combines design, component, manufacturing, standards and specifications, cost data, and technical information into one system. Such a product model needs to

- support simultaneous collaboration of multiple users while possessing some method for configuration control of changes
- provide visualization and "walk-through" capabilities so that designers can determine whether a given design can be manufactured and supported
- link to manufacturing and support databases/equipment.

Such tools help to reduce problems that traditionally were discovered only during manufacturing. However, they are very costly to develop and maintain. For the IPPD process where there is a rapid pace of design and refinement—as well as a broad range of design maturity—such a system is almost required.

Development of such a system can be a significant undertaking. No off-the-shelf tool works "out of the box." Some customization of the tool will be necessary to accommodate the specific shipbuilding process, manufacturing environment, and business processes. Greater numbers of users and more complicated designs require more sophisticated systems. Furthermore, if users are located at two or more sites, then such issues as data transfer and security must be considered.

Private Sector Versus Government

As discussed above in the section on roles and responsibilities, the government or industry could lead or conduct these activities. A government with sufficient technical resources can quite effectively lead the early conceptual phases when many design iterations and options are considered. Very few countries have the technical resources to conduct

---

8 Adapted from Birkler et al., 2011.

detailed design. In fact, it is better to have the builder and designer closely linked so that producibility issues are addressed during design (see the section above on IPPD).

Readers should note that in this chapter we primarily addresses the design phases associated with the HM&E systems of a new ship class. There may be similar design efforts conducted in parallel for the major combat and weapon systems that will be installed on the ship. The various design teams must coordinate to ensure that the integration of the systems with the basic ship goes smoothly. Agreed upon interfaces and boundaries must be established during the design process that define the connections between the combat and weapon systems and the ship. Also, sufficient power and cooling must be available to support the various systems.

Design-Related Factors the Program Can Control

There are some external factors over which the program has little or no control, but it can control a number of factors and decisions. First, as noted in the chapters above, it is important to involve all appropriate organizations in designing the ship. Although a pure MOTS acquisition option will provide a proven design, an evolved MOTS or new design option requires either modifications to an existing design or the development of an entirely new one. For the evolved MOTS and new design options, it is important to have builders, maintainers, operators, and the technical community involved in the design process.

An important lesson from the *Virginia* program is to use a design-build process during the design of a new combatant. This involves having the builders actively involved in the design process to ensure that what is designed can be built in an efficient manner. Design-build should go further than merely involving builders in the design process. The design should also be informed by operators, key suppliers, maintainers, and the technical community. Therefore, it is important to think of the design team as a collaboration of draftsmen and design engineers with inputs from those who must build to the design, operate the ship, and maintain it. This collaboration should extend throughout the design process. However, throughout that process, it is important to keep in mind that the ultimate design and construction target is a ship that is cost-effective in its postdelivery and in-service phase. Although maintenance ease is a desired trait, it must be balanced against long-term maintenance costs.

The program also needs to listen and react to the concerns that the technical community may raise. The degree to which existing technology is "pushed" in a new design will affect the risks to cost, schedule, and performance of the platform. The technical community must understand the state of technology and the degree to which a new design extends that technology.

The technical community consulted during a new design effort should extend beyond the in-country resources to include the technical assets of partner nations. In some areas, especially technical ones not encompassed in previous programs, other countries may have a deeper and better understanding of the technology and risks. For example, the Australian technical community may have knowledge of a specific combat system but very limited experience with it.

Ensure That the Selected Design Organization Understands the Concept of Operations and Build Environment

If the SEA 5000 decisionmakers select a MOTS or an evolved MOTS option, it will need to collaborate with a design house. Shipbuilding programs have typically been more successful when the same organization is contracted to design and to build the vessel. In the past, because of Australia's lack of a domestic design house, previous ship classes have been designed and built by separate organizations through partnership agreements. If this is the approach taken, the program office will have to carefully manage these organizations to ensure a smooth transition from design to build.

It is imperative that the design organization understand and appreciate the ship's concept of operations and operating environment. Operational environments differ (i.e., weather patterns, sea states, salinity) around the world; therefore, the design considerations for the RAN will differ from those of some European navies. For example, in the *Collins* program, the selected design organization—a Swedish design house—was accustomed to designing ships for the relatively calm, cold waters of the Baltic. The *Collins*, on the other hand, would primarily be operating in an open-ocean, tropical environment. The operational and support requirements for this type of environment were not well understood by the designer, which eventually led to equipment and system problems during the construction and operational life-cycle of the class.[9]

The designer also needs to take into consideration the build environment. For example, caution should be taken when adapting designs intended for a single-builder production process to a distributed or modular build strategy. The Australian National Audit Office found that for the AWD program, the risks associated with executing Navantia's F-104 design in Australia's distributed-build project were underestimated and not fully realized until well into the build phase of the program.[10]

Specify Adequate Design Margins and Manage Them During the Design and Build Program

A general lesson throughout most ship design and production programs is that a new ship design must include adequate weight, stability, power, cooling, and bandwidth margins that must be closely managed during the design, build, and operation of the ship.[11] New ships and submarines typically start with what are believed to be adequate design margins, but they are often consumed during the design and build process or early in the platform's life. Without adequate margins, it may not be possible to modernize and upgrade equipment. New power and cooling plants may be needed, but they may exceed available weight margins. Existing systems may be downgraded or ship operations may be constrained if adequate margins are not available.

When using and modifying an existing MOTS platform there may already be less of a design margin than desired. For example, modification of the DDG-51 design over time has used up some of

---

[9] Schank et al., 2011a, p. 22.

[10] Australian National Audit Office, 2014, p. 35.

[11] For a further discussion of design margins, see Chapter Six.

the original design's growth margin. The third flight (Flight III) of the DDG-51 would have less of a growth margin than what the Navy would aim to include in a new destroyer design of the same size. This has driven concerns that the Flight III would be limited in the amount of additional capability that could be achieved over the platform's projected service life.[12]

It is difficult to estimate the potential cost effects of greater design margins, but some analysis of ship cost models suggests that a more flexible architecture in design will have slightly higher costs for the lead ship of the class but will also provide the potential for significant cost savings over the entire class of ship. Table 7.2 compares acquisition costs for the DDG-51 (an inflexible platform design) to a notional flexible platform. Analysis also suggests that more flexible platforms also achieve greater capability at the end of service life.[13]

Include in the Design the Capability to Remove and Replace Equipment

The operational life of a ship is typically greater than the life of some of the technologies incorporated in its design. This is especially true for C4I equipment. The design should include adequate access paths and removal hatches to facilitate removing and replacing damaged or

Table 7.2
Procurement Cost Comparisons ($FY10, millions)

	Inflexible Platform (DDG 51)			Flexible Platform (SCAMP)		
	Lead	Average	Class	Lead	Average	Class
Procurement	1,700	1,260	92,280	1,380	850	62,500
R&D	1,130	50	4,550	1,530	20	3,070
Acquisition	2,830	1,310	96,830	2,910	870	65,570

SOURCE: Page, 2011.

[12] Ronald O'Rourke, *Navy DDG-51 and DDG-1000 Destroyer Programs: Background and Issues for Congress,* Washington, D.C.: Congressional Research Service, RL32109, April 8, 2014b.

[13] Page, 2011, p. 55.

obsolete equipment. For C4I equipment, modularity and interoperability should be incorporated into the design.[14] Data and information architectures should be developed that allow the installation of electronic equipment as late in the build process as possible to take advantage of rapid changes in information technology. Open architectures should prove useful to equipment integration and future modernization efforts.

Understand the Technical and Integration Risks

As the technical complexity of naval ships increases, it creates challenges for the integration of new systems. Revolutionary technologies inject high levels of risk into a program. However, the rapid pace of technology change, particularly for software-dependent systems, means that from the time the ship enters development until the time it is delivered, the state-of-the-art technology may be generations removed from the initial concept. Meanwhile, the push to meet in-service targets and to move ships forward into the construction phase often leads decisionmakers to do so under situations of unstable design or immature technology.

The program needs to have an understanding of the level of technology and integration risk in the design, particularly for critical technologies.[15] Best practices suggest that for critical technologies, technology readiness level (TRL) 7 is the level of maturity that should be achieved by the time the contract is awarded for detailed design. The integration phase for combat systems software is often the most challenging, requiring specialized skills and facilities. Thus, software

---

[14] For a discussion of controlling the C4I upgrade costs on ships, see John F. Schank, Christopher G. Pernin, Mark V. Arena, Carter C. Price, and Susan K. Woodward, *Controlling the Cost of C4I Upgrades on Naval Ships,* Santa Monica, Calif.: RAND Corporation, MG-907-NAVY, 2009.

[15] A technology element is defined as critical if "the system being acquired depends on this technology element to meet operational requirements and if the technology element or its application is either new or novel or in an area that poses major technological risk during detailed design or demonstration." See GAO, *Best Practices: High Levels of Knowledge at Key Points Differentiate Commercial Shipbuilding from Navy Shipbuilding,* Washington D.C.: U.S. Government Printing Office, May 2009.

integration has a high risk for cost and schedule delays, even for existing platforms. For example, in 2012, the U.S. GAO found that integration issues with the multimission signal processor for the DDG-51 Flight III set the program four months behind schedule; the slippage produced some $10 million in realized cost growth with an additional $5 million projected.[16]

Manufacturing readiness levels are also an important factor to consider in the evaluation of technology risk. Best practices suggest that the manufacturing processes for new systems should be tested and validated, ideally through prototype manufacture, before the initiation of production. In the DDG-1000 program, the revolutionary design for a composite deckhouse—a critical technology intended to reduce weight—required the development of new manufacturing and assembly processes. Although the shipbuilder validated these processes through the development of a large-scale prototype, the test and inspection activities and required facility and machinery upgrades were not completed before production readiness review and initiation of construction.[17] This led to cost and schedule implications for the program, and the government eventually abandoned the composite deckhouse and returned to a steel design.

The development of the Ashvale development facility and the Warspite facility at the Barrow shipyard were positive lessons in technology and manufacturing readiness from the *Astute* program. These facilities provided for early testing of the *Astute* combat system before it was installed on the submarine. This helped to optimize the production process and to reduce the risks of combat system operations and integration. The command-and-control off-hull assembly and test site (COATS) facility at General Dynamics Electric Boat performs a similar function for *Virginia*-class submarines by allowing the command-and-control system to be assembled and tested before it is inserted into the submarine.

---

[16] GAO, January 2012a, p. 26.

[17] GAO, *Defense Acquisitions; Zumwalt-Class Destroyer Program Emblematic of Challenges Facing Navy Shipbuilding,* Washington D.C.: U.S. Government Printing Office, July 2008a, p. 5.

Develop an Integrated Master Plan for Design and Build

A program should have an overall integrated master plan detailing the tasks, milestones, and products that are expected during the design and build of the ship. The integrated master plan should describe the order of tasks, events, and interrelationships. It should define the critical path for achieving schedule targets and should highlight the potential effect of delays. The integrated master plan should be systematically monitored and proactively managed to ensure that tasks remain on schedule. Schedule delays are often seen as an indicator of program performance issues and may trigger additional oversight, reviews, or certifications. Given the pressure on program managers to meet scheduled milestones, it is important that schedule targets are both realistic and executable.

The GAO has outlined ten best practices for developing a realistic and executable program schedule; these are shown in Appendix C. These best practices emphasize the importance of building in reasonable float (slack) into the schedule but at the same time maintaining tighter targets for activities that are part of the critical path. In addition, particular attention should be given to activities in the critical path that are "high risk," including the development of immature technologies or complex integration

Consider Potential Problems with Foreign Suppliers

With a pure or evolved MOTS option, it is likely that various equipment and systems will be provided by non-Australian companies. The *Collins* program relied on foreign suppliers for key equipment. Often, a lead item would be built in another country and then production drawings would be provided to an Australian company to build the remaining items. On the surface, this transfer of build processes should work, but there were examples where the "tribal knowledge" of the build procedures was not addressed solely by the construction drawings and plans. The electric generators, designed by a French company, are a prime example of such problems: The Australian company lacked the knowledge or specialty manufacturing equipment or systems required to build them. The Hedemora engines are another example of a foreign supplier not being able to adequately address problems that emerged. If foreign suppliers are chosen for key equipment in a new program, they need to provide assurances that they are economically viable and will remain so during the operational life of the ships. Also, if equipment designed by a foreign organization is to be built in Australia, personnel from the foreign supplier should interact, preferably in Australia, with the company building the equipment to provide the detailed knowledge needed beyond that captured on design drawings.

Supplier issues go beyond companies outside Australia. An adequate supplier network inside Australia must also be developed and nurtured to ensure that the vendor base exists when needed. Maintaining an adequate vendor base is the responsibility of both the government and the shipbuilder, since some parts and equipment are bought and provided to a program by both parties.

Key Points for the Design Phase

- *Ensure that the selected design organization understands the concept of operations and build environment.* This is particularly important when selecting a foreign design house that may not have the necessary context for Australia's operating environment and industrial capabilities.
- *Specify adequate design margins and manage them during the design and build program.* In-service upgrades to ship equipment and systems typically require additional power, cooling, bandwidth, and other ship services. Without adequate margins, it may not be possible to modernize and upgrade equipment.
- *Include in the design the capability to remove and replace equipment.* Upgrading and modernizing a ship's equipment can be costly and time-consuming. Cost and schedules can be reduced if equipment is easily accessible.
- *Understand the technical and integration risks associated with the ship design.* Pushing technologies to a large degree or in many areas leads to cost and schedule risks. Technologies should be well developed before entering into a design and build contract.

- *Develop an integrated master plan for design and construction.* The master plan should identify key milestones and critical paths to achieving schedule goals.
- *Consider potential problems with foreign suppliers.* An adequate supplier network and communication processes should be put in place to effectively manage issues with foreign vendors.

CHAPTER EIGHT

Manufacturing and Build

This chapter describes major events and important considerations during the construction of the ship. It discusses the various resources required during ship construction, the establishment of a supplier base, the use of advanced outfitting practices, outsourcing, and the use of multiple shipyards in the construction of the ships. The chapter concludes with the choice between GFE and CFE and the various metrics that are used to track progress during the construction phase.

Definition and Activities

The process of modern ship construction has evolved over the past few decades. Traditionally, ships were built from the bottom up (also referred to as stick-built), starting from the keel and building upward in a shipway or building dock. The ship was structurally supported until the hull was complete enough to support itself. Parts, equipment, and components were brought to the shipway and installed. Such a building approach has limitations in terms of efficiency. One limit is that most of the work was exposed to the elements, so weather could dramatically affect productivity. Also, workers did construction work in confined and sometimes difficult to access areas. The logistics of moving equipment, tools, and material into the hull was time-consuming. Last, the orientation of the structures was always vertical, so working on overhead parts of the ships meant unnatural ergonomics for the craftsman.

Modern shipbuilding overcomes these limitations by using a construction technique that is piecewise, or more properly termed "modular," construction. Self-supporting sections of the ship are built inside covered work areas or near the assembly dock. Modular construction is typically described in terms of the various sizes of work items. Definitions of these pieces that make up a ship include[1]

- Structural Unit: A three-dimensional structural assembly whose dimensions are usually driven by the maximum plate or panel line size, which has all welding complete and contains varying degrees of outfitting.
- Block: A structural part of the ship's hull consisting of plates and reinforcing frames, generally produced by erecting and joining together panels, assemblies, subassemblies, units, and parts. These parts can be erected on the ship as a block or combined with other blocks and units to form a grand block.
- Grand Block: An assembly of blocks that may be built in a fabrication facility or on an outside platen area. Grand blocks usually involve large-capacity cranes or transporters to move and lift them into the assembly dock, shipway, or land-level facility.[2]
- Packaged Unit or Module: A grouping of outfit items installed on a common foundation, such as a machinery packaged unit or piping unit, before installation on a block, grand block, or assembled ship.

[1] Different shipbuilders use different terms for the various pieces that are formed during ship construction. The lowest-level structural units are also called *assemblies* or *modules*. Grand blocks are sometimes referred to as *superlifts*. There are also differences related to the size of the piece. For example, what some shipbuilders refer to as *blocks* may be called *grand blocks* by other shipbuilders. These differences in terminology caused shipbuilders some problems in responding to the survey and caused us some problems in analyzing their responses.

[2] Ships may also be constructed from super blocks, which are large portions of a ship's hull or a deckhouse made up from blocks and grand blocks. Australia's AWD and LHD programs and the UK's Type 45 and CVF programs have built their ships from several blocks constructed at various shipyards and transported to and assembled at one shipyard.

- Assembled Ship: The joining of blocks and grand blocks to form the "complete" ship, typically done in a dry dock, shipway, or land-level facility.

The size of these assemblies, blocks, and grand blocks is highly dependent on the lift infrastructure of the fabrication yard.

Changing the hull construction method improves productivity and lowers costs by

- reducing the man-hours to complete the hull
- reducing associated overhead burden (through reduced assembly facilities)
- increasing revenue by means of faster construction.

This increase in efficiency occurs because the objective of modular construction is to complete as much work as possible at an early stage. This can be done because the work is

- executed under cover, in a good environment with no weather disruptions
- accessible, with little or no staging or other access equipment
- performed downhand, so workers are comfortable and effective
- conducted with automation applied as far as possible to reduce costs.

The difference in man-hours spent in different assembly, block, grand block, and ship stages of production is significant and is discussed in more detail below.

Required Resources

Personnel—Skills

Similar to the technical workforce, there are a great variety of manufacturing skills that constitute a shipbuilding enterprise. We have broken these skills into three broad categories: structure, outfitting, and direct support. Table 8.1 lists the various subskills within each of these categories.

Facilities[3]

The production of a ship or submarine involves numerous facilities, including a wide range of shops, cranes, specialized equipment, docks,

Table 8.1
Manufacturing Skills in Warship Construction

Skill Category	Subskills
Structure	Steelworker, plater, boilermaker
	Structure welder
	Shipwright/fitter
	Team leader, foreman, supervisor progress control (fabrication)
Outfitting	Electrician, electrical technician, calibrator, instrument technician
	Hull insulator
	Joiner, carpenter
	Fiberglass reinforced pipe laminator
	Machinist, mechanical fitter/technician, fitter, turner
	Painter, caulker
	Pipe welder
	Piping/machinery insulator
	Sheet metal
	Team leader, foreman, supervisor progress control (outfitting)
Direct support	Rigger, stager, slingers, crane and lorry operators
	Service support, cleaners, trade assistant
	Stores, material control
	Quality assurance/control

SOURCE: Arena et al., 2005b.

[3] From Arena et al., 2005b.

and piers. Producers employ facilities at different times, in different sequences, and in different ways, depending on the platform being built, yard organization or layout, build strategy, and many other factors. Table 8.2 depicts the three production stages (pre–final assembly, final assembly [FA], and afloat outfitting [AO]) with corresponding facilities each requires. These stages are activities that occur during the construction shipbuilding phase (described above).

The pre–final assembly phase entails a manufacturing period before final assembly of blocks and modules begins, that is, the period before the ship occupies an assembly location. During this time, such facilities as pipe fabrication shops, unit assembly areas, laydown areas, and steel fabrication shops are used. Final assembly begins when a producer starts assembling the ship using a facility such as a dry dock, floating dock, slipway, land-level area, or ship assembly hall. Afloat outfitting begins when a ship is launched or floated and ends when the ship is delivered. A ship in the AO phase would require a pier, quay, lock, or a dock.

There is some overlap in use of different facilities throughout each phase. In many cases, certain facilities—cranes, shops, or fabrication facilities associated with the pre–final assembly—are used throughout

Table 8.2
Facilities Use During Ship Production

Production Stage	Facilities Used
Pre–final assembly	Shops Cranes Module halls
Final assembly	Dry docks Cranes Floating dry docks Slipways Land-level areas Ship assembly halls
AO	Cranes Piers Quays Locks

SOURCE: Arena et al., 2005b.

the FA and AO phases. Generally, the FA and AO phases are mutually exclusive, but sometimes an FA facility will be used for outfitting.

Establishing a Supplier Base[4]

The establishment and qualification of a new vendor base can be a very time-consuming activity for the prime/lead contractor. This level of effort will depend on the design choice by DMO/RAN. A MOTS design will likely leverage an existing vendor base, so there might be minimal efforts to validate earlier vendor quality or replace vendors who have left the market. With a clean-sheet design, qualification will take place for all the required items.

The qualification process can be thought of as having five steps, as laid out in Figure 8.1.

The full qualification process includes the vendor qualification, design qualification, first article testing, repeat process testing, and audits. As mentioned, not all of these steps will be included in every qualification process.

Vendor Qualification

Vendors that are working for a prime for the first time must go through a qualification process. Vendor qualification is done by the primes and takes place before a vendor is allowed to bid on a contract. The prime looks over the accounting systems, managerial structure, prior experience, and technical competency of the firm.

The process is rather straightforward, with the prime contractor typically providing descriptions of the various types of data needed

Figure 8.1
Vendor Qualification Process

RAND *RR767-8.1*

---

4 Section adapted from Schank et al., forthcoming.

from the vendor. The required information is often readily available from existing data and rarely requires new data-collection efforts. Prime contractors will also periodically monitor the health of a vendor to hopefully identify any problems that could occur in the future.

Design Qualification

Once qualified, a vendor can bid on contracts. The prime contractor defines the requirements and technical specifications for the product as well as the testing process to ensure that the product meets the requirements. The vendor will then provide its design, manufacturing process, and test plan to the prime contractor for initial approval. The test plan will provide details on the testing process, including how testing will be conducted and evaluated. The prime contractor will check the design deliverables and product schedules to ensure that they meet production plans. The prime contractor and the vendor will iterate the proposal until it is ready to send to the government for its approval. The prime contractor and government will ensure that the vendor has the facilities and expertise to build to the approved design. The government will supply comments and work with the vendor and prime until the government approves the design. Certification organizations, such as the American Bureau of Shipping or Lloyd's Registers, also have a role in the design qualification process. For example, a certification organization may specify specific welding procedures and worker qualifications.

The design qualification process can be costly for both the vendor and the prime contractor. The vendor must have the expertise to design the product and produce the manufacturing and test plans. The prime contractor may have to invest significant time and energy interacting with the vendors, especially small or new vendors, to help them work through the requirements and testing processes. A major shipyard/prime may have hundreds of personnel who oversee the supplier chain and work with vendors to ensure that their products are qualified.

The process for qualifying the design of the product is greatly simplified if the government or vendor holds technical data rights to an existing design. The design does not require qualification approval. The vendor has to demonstrate the ability to adequately manufacture the design and test the final product.

First Article Testing

The first article testing includes the detailed design of the product and prototype manufacturing in addition to the test process itself. Before the first article can be tested, the test procedure must be approved by either the prime or the government depending on the part. Vendors may not have adequate facilities and capabilities to test the product and often use independent facilities for the various steps in the testing process. Government facilities are typically more expensive. The prime contractor, the government, or an independent verification organization may witness the testing process and make a judgment on the results.

The types of tests performed vary by weapon system or product but can include shock, vibration, electro-magnetic interference (EMI), or others. Generally, aircraft require fewer tests than ships. Most parts for ships require the full range of tests. Major equipment requires more testing than such commodities as steel, cable, valves, or various fittings. Also, electronic equipment typically requires more testing than hydraulic equipment.

Repeat Process Testing

This is done to ensure that the quality of manufactured products remains compliant with the requirements and that the vendor can produce the product in the quantities desired.

Audits

Once manufacturing has begun, the government client will occasionally audit the vendors to ensure that they continue to meet the required standards.

Important Considerations

Design Maturity at the Start of Construction

An important management decision facing naval shipbuilding programs is deciding the point in the design process at which to begin construction. This decision is a compromise between achieving design stability (and lowering subsequent construction risk) and maintaining capability in the industrial base. A best practice on defense programs, according to the GAO, is to have at least 90 percent of the design drawings complete before beginning production and all the technical and stakeholder reviews complete.[5] The GAO further refined this view for shipbuilding programs, such that the appropriate starting point is when "basic and functional design" is complete before the start of construction—meaning that the 3-D product model is complete (with final outfitting details and vendor information); however, all production drawings and instructions need not be complete.[6] Thus, in ideal circumstances, starting production as late in design as possible reduces risk.

However, naval shipbuilding programs also face risks from the loss of workforce skills and capabilities in the industrial base. It is very rare that multiple programs will be running concurrently, so that while a new class is in design, an older one is being produced. Generally when a new design begins, production workload starts to taper off—placing at risk the ability to maintain highly skilled workers. The longer the gap, the more risk there is to the workforce. Therefore, industrial base considerations will drive a construction start as early as possible to keep the workforce actively employed. So, practically, there will be a compromise between design and workforce risk in choosing the start point of construction.

Advanced Outfitting[7]

Outfitting tasks occur either during the construction of the pieces that make up the ship or when those pieces are assembled to form the

---

[5] GAO, *Best Practices: Capturing Design and Manufacturing Knowledge Early Improves Acquisition Outcomes*, Washington, D.C.: U.S. Government Printing Office, GAO-02-701, July 2002.

[6] GAO, 2009.

[7] The following was adapted from John F. Schank, Hans Pung, Gordon T. Lee, Mark V. Arena, and John Birkler, *Outfitting and Outsourcing Practices: Implications for the Ministry of Defence Shipbuilding Programs,* Santa Monica, Calif.: RAND Corporation MG-198-MOD, 2005a.

completed ship. Outfitting covers a broad range of functional tasks including:

- Structural: Installing equipment foundations, doors, ladders, hatches, and windows.
- Piping: Installing and welding pipes, including spools and connectors.
- Electrical Power Distribution: Installing the power distribution system downstream of the main power switchboards, including hanging and pulling cables and installing local switchboards and ancillary electrical equipment.
- HVAC: Installing air handling units, ducting, and other ancillary HVAC equipment.
- Joinery: Installing accommodations, such as cabins or berths, dining facilities, food preparation areas, and rooms for meetings or other administrative purposes.
- Painting and Insulation: Covering the structure and accommodations of the ship.

For naval combatants, outfitting also includes the installation of combat and weapon systems.

One advantage of modular construction is that it allows for advanced or early outfitting. Advanced outfitting involves performing those outfitting tasks early in the ship construction process, i.e., at the unit, block, or grand block stages. Advanced outfitting allows the outfitting tasks to be accomplished in covered production facilities or on nearby staging areas where the material and equipment are close at hand and where the workforce and construction units are protected from adverse weather. Also, performing outfitting tasks in production facilities allows the structural elements to be positioned in the best way to allow easier installation of material and equipment.

Early outfitting reduces the labor time and cost. Often, rules of thumb are quoted, such as "1-3-5-10," indicating the number of hours to perform a given task at the unit stage (1), the block stage (3), the grand block stage (5), or on the assembled ship (10). The format of these rules of thumb can vary among shipbuilders. For example, some shipbuilders may have only three metrics in their rules—for block, grand block, and the assembled ship. Other shipbuilders will add a metric for ships that have been launched. Regardless of the format or the specific metrics, all shipbuilders commonly believe there are savings from advanced outfitting. However, few published data exist to measure the actual time and cost effects.

In addition to reducing labor hours, shipyards may strive for higher levels of early outfitting to reduce the time spent in constrained facilities during ship construction. For example, at many shipyards, the erection dock is the bottleneck in the sequential construction of ships. Reducing the time a specific ship spends in the dock allows the shipbuilder to begin construction of follow-on ships sooner. Transferring outfitting hours to the shops or to the assembly areas alongside the dock reduces the hours spent in the dock, thus enabling higher capacity utilization, and therefore, higher productivity, even if there is no overall reduction in the number of hours to build a ship.

Almost all shipbuilders believe that the lack of timely design information affects their production planning and management process and adversely affects the degree of advanced outfitting they can accomplish.[8] Most also feel that delays in delivery of outfitting material and equipment, caused by either incomplete designs or delayed contracting for long-lead items, limited their ability to accomplish higher levels of advanced outfit. Concern for damage of outfit equipment, especially in the joinery area, and limitations imposed by the customer can also be major factors limiting the degree of advanced outfitting that can be accomplished. The time required to resolve questions or issues on the clarification of requirements that arose during the construction of the ship was a factor in this area.

---

[8] Some U.S. shipyards are involved in long production runs where basic designs are well established, thereby facilitating advanced outfitting.

Outsourcing[9]

Outsourcing occurs when shipbuilders subcontract certain work during ship construction to other firms, either other shipbuilders or to nonshipbuilding organizations. Such an approach can succeed where there is a robust shipbuilding industry to support such activities. Outsourcing can be used in two very different ways: total outsourcing and peak outsourcing.

Total outsourcing involves a shipbuilder subcontracting a complete functional task, such as electrical, HVAC, or painting, to an outside firm. In this case, the shipbuilder retains no in-house labor capability to perform the function although the shipyard may provide facilities (such as painting sheds) or materiel and equipment to the subcontractor. The subcontractors may be turnkey, where they provide the design and the materiel and equipment and perform all installation, or partial turnkey, where the shipbuilder may provide some combination of the design, facilities, and materiel and equipment.

Peak outsourcing occurs when a shipbuilder uses a subcontractor or temporary labor to augment in-house capabilities during times of peak demands or to accelerate operations when schedules start to slip. Many shipyards also use peak outsourcing to adjust workforces when demands decrease in light of strict national labor policies restricting the termination of permanent employees. The subcontractors may work at the shipyard alongside shipyard employees or at their own sites. An example of the latter is having an outside firm build portions of the ship structure and send them to the shipbuilder for integration into the final ship. Peak outsourcing may be used for all ships that a shipyard builds or for only certain ships in its product line.

Outsourcing may offer several advantages to the shipbuilders involved in the construction program and to the program itself. These include:

- Alleviate shipbuilder workforce shortfalls: The shipbuilder may have shortfall in certain skill specialties that would require hiring

[9] This material was adapted from Schank et al., 2005a.

workers with low proficiencies and training those workers in the skills and proficiencies needed.

- Reduce construction cost: Potential savings are due to reduced overheads, potentially lower wage rates, lower costs associated with the hiring and dismissal of shipyard workers to meet cyclic demands, and improved quality that can lead to fewer man-hours and less rework associated with certain construction tasks.
- Reduce the need for new capital investments: New ship work may result in the need to modify existing facilities or construct new facilities at shipyards involved in the program. The use of subcontracts may negate the need for these facility enhancements, thereby reducing capital investment costs.
- Assure uniform quality of ship systems on modules: Build plans may call for several large sections or modules to be built in various shipyards and then transported to one shipyard for final assembly and test. Each module will be self-contained in terms of electrical power distribution, piping systems, HVAC systems, and accommodations. Common subcontracting firms used by all shipyards building modules may provide uniform systems for the various super blocks.
- Ensure compatibility between systems that go across modules: Some outfitting work will be done on the assembled ship rather than on each super block. For example, splicing cable is often prohibited on military and commercial ships. In those cases, cable is installed after the final assembly of all the blocks and modules. Also, crew accommodations and other "hotel" functions—such as food service, laundry, and waste disposal—are installed on assembled commercial ships to eliminate potential damage during the construction of the blocks. Again, using a subcontractor for those systems that go across the super blocks may ensure uniform quality.

Multiple Shipyards[10]

Many commercial and military ships have been assembled at one shipyard from modules built at multiple shipyards. Some recent examples of a shared modular build approach for naval ships include:

- U.S. Navy DDG-51 destroyer deckhouse
- U.S. Navy DDG-1000 destroyer
- U.S. Navy *Virginia*-class attack submarine
- UK's Type 45 destroyer
- UK's *Queen Elizabeth* aircraft carriers (QEC)
- France's *Mistral*
- Australia's AWD
- Australia's LHD program.

The reasons for adopting such an approach are diverse and include providing work to more than one geographic region, maintaining a shipbuilding industrial base, accessing skills available only at different shipyards, overcoming capacity constraints, and reducing costs.

However, such an approach involves many challenges. The client is often involved in decisions regarding how workload gets allocated between the shipyards, which can often have a political dimension. Another challenge is how to set up the contractual arrangements with the shipyard. Examples of some of the strategies employed are an alliance, prime-subcontractor, and government to contractor (e.g., government supplies ship modules as GFE to the assembler).

Another challenge is how to coordinate and control technical information between the shipbuilders. Very strict control of the module interfaces is vital to a successful shared-build program. Also, the shipbuilders have to use the same design tools and product models to easily exchange information. For example during the *Virginia*-class program shared build, General Dynamics Electric Boat and Huntington Ingalls Industries Newport News Shipbuilding established termi-

---

[10] The following was adapted from Laurence Smallman, Hanlin Tang, John F. Schank, and Stephanie Pezard, *Shared Modular Build of Warships: How a Shared Build Can Support Future Shipbuilding*, Santa Monica, Calif.: RAND Corporation, TR-852-NAVY, 2011.

nals at Newport News and tied them to Electric Boat's product model so that the two companies could readily exchange technical details in the same format.

Risk reduction is another aspect of relying on multiple shipyards. Although it is difficult to directly compare different programs, there are four areas in which risk reduction is important: motivating cooperation, design completion, design and design-to-production organization, and aligning production practices and schedules. Failure in the earlier stages of this progression is more likely to be catastrophic to the program than later on. The importance of these risk reduction areas increases with the complexity of the modular build process, which, in turn, is linked to the complexity of the vessel. In other words, shared-build risks are likely much higher for more complex vessels than traditional build risks for such vessels. Reducing risk will give greater assurance that a program can deliver the required vessels, that they will be delivered on time, that they will be of the required quality, and that the program will meet its cost targets.

Contractual obligations and financial remuneration are only part of the process to motivate shipyards to cooperate. At the heart of cooperation is the trust between the yards at almost every level of management and production. Trust can be improved with open book accounting and a fair division of profits or losses. The prospect of future competition might impede open-book accounting and the degree of trust that can be achieved. Yards are unlikely to share best practices if there are proprietary concerns. The government or the Navy needs to encourage cooperation between reluctant shipyards; this might be done through contracting structures, workload allocation, or any other lever available to government.

A modular-build process requires greater detailed design completion before construction starts than traditional stick build shipbuilding. Detailed design is a key step in mitigating rework requirements by allowing better quality control and ensuring accurate and timely stock delivery to the production process. This becomes even more important when modules are built at two or more locations. In particular, the design at the interfaces of the modules needs to be fully understood and, therefore, practically complete for modules to integrate easily.

Additionally, producing complex modules that are easy to integrate requires that the yards reach a detailed and common understanding of what affects module interfaces. This includes the part manifest and other aspects of production linked to detailed design. The involved yards must detail the build assignments down to the piece-part level at the interfaces. To achieve such commonality requires design software that is either common or linked to a common design data bank. For complex designs, IT systems and the supporting software can become very complex and susceptible to instability.

Modular construction within a single yard is easily aligned, but that between two yards, particularly those that have not worked together or that are at different stages of modernization, can be difficult to achieve. Aligning production practices requires that each yard, in particular the integration yard, understand differences in production processes. This is of vital importance at the interfaces of complex, outfitted modules. Many integration yards will send their own personnel to implement additional quality assurance processes at the supplying yard. Aligning the production schedules also requires pacing module construction to the same completion drumbeat. If the planned work on a block in one yard becomes late, it might be necessary to deliver that block unfinished to the other shipyard. It might then be finished by a deployed team from the originating yard or by labor at the receiving yard. Such an arrangement will affect subcontractors and suppliers and likely add further to costs. Alternatively, the block could be delayed until complete, with potentially serious scheduling effects on the production plan at the receiving yard.

Whether building at multiple shipyards saves money depends on the circumstances. There are definitely additional costs associated with a multiple-yard build compared with a single-yard build. These include transportation, module and facility infrastructure, IT, and additional quality assurance and oversight. Savings might result for many reasons. For example, one shipyard might be more efficient at building a certain section of ship. Or it might be less expensive to use multiple yards rather than to expand the workforce at a single yard. But in the end, multiyard programs always have additional costs and it is not clear that they always generate savings.

Gap Between First and Second Hull

A shipyard is most efficient when its workload is consistent so that it can employ a stable workforce (does not have to go through cyclic lay-offs or hiring). The challenge for most shipbuilding programs is the desire to have some sort of gap between the first and second hull of the class. This allows the program to integrate learning and design changes from the first ship onto the subsequent hulls. But a long gap becomes impractical from a workforce standpoint. Often the second hull must begin before the first hull is complete and tested. The build frequency must be a compromise between design maturity and industrial stability.

GFE Versus CFE Decisions

Either the government, the shipbuilder, or a contractor can purchase the major equipment needed for a ship. Typically, when the government is controlling the features and specifications (or is procuring through another prime), it takes responsibility for procuring and furnishing equipment (GFE). Weapon systems are examples of GFE items on naval ships. Such items may be common across several platforms and purchased on a common contract. One risk with GFE is that the government has the responsibility to deliver both technical information and hardware in a timely manner so as to not disrupt the design-build process.

Contractors can more effectively procure other material (CFE). Contractors typically procure commonly available HM&E equipment, for example. In U.S. procurement, the preference is to have the contractor be responsible for material items that may affect contract performance. So equipment items whose delivery might be on the critical path are not good candidates for GFE. GFE also requires additional government resources to manage.

There are advantages and disadvantages to both GFE and CFE-systems.[11] GFE equipment allows standardization across the fleet, and such equipment is easier to support once a ship enters the fleet. Complete standardization across the fleet of even one system is probably an unattainable goal, however, because of the refresh rates of some

[11] This paragraph adapted from Schank et al., 2009.

technologies and the varying timelines by which ships are available for upgrades. CFE systems can leverage the broader commercial marketplace and provide multiple technology options, hopefully at a lower cost.

However, CFE systems can present a problem to the RAN once a ship enters the fleet and responsibility for the support of the systems transitions to the operators, who may not have complete knowledge of the systems. Also, CFE systems may create a unique logistics tail, requiring a separate spare parts pool and system-unique training for the sailors who operate and maintain the system. For example, this issue was a major problem with the Shipboard Wide Area Network developed by the shipbuilder for the LPD-17 class of ships for the U.S. Navy. Once those ships entered the fleet, the ship support elements had difficulty understanding the nuances of the system and, therefore, had difficulty upgrading and supporting the system.

Tracking Progress[12]

Cost and schedule control and estimating are central competencies of program management. Continually updated knowledge of project status is important for both operational planning (i.e., determining when the customer will have use of the asset) and financial management (i.e., determining the cash flow needed to support the program). A good control system can also aid program improvement by identifying problem areas before they greatly affect production. Accurate estimating of changing program needs allows an organization to make best use of limited funding.

In an earlier RAND report,[13] we surveyed major shipbuilders in Europe, the UK, and the United States and conducted follow-up, in-depth interviews with representatives of these firms. From those surveys and discussions, we identified metrics that are most commonly

---

[12] Adapted from Mark V. Arena, John Birkler, John F. Schank, Jessie Riposo, and Clifford A. Grammich, *Monitoring the Progress of Shipbuilding Programs: How Can the Defence Procurement Agency More Accurately Monitor Progress?* Santa Monica, Calif.: RAND Corporation MG-235-MOD, 2005a.

[13] Arena et al., 2005a.

used to track shipbuilding progress. These metrics fall into six general categories: earned value–related, milestones, task-oriented, actual versus planned, area/zone (such as compartment completion), and miscellaneous. We asked the shipbuilders to report their primary schedule control metric during each of the five phases of shipbuilding: design, module block construction, assembly, outfitting, testing/trials, and commissioning.

Table 8.3 reports the fraction of shipbuilders using a particular metric during each phase for the six major metric areas. In each box, we put the metric category at the appropriate fraction. Within each box, the metrics are ordered according to increasing fraction (higher at the top). Earned value metrics are most commonly used in each design and construction, though less frequently in test and trials. Milestones are the second most commonly used tracking metric in design and construction.

The preference by region (U.S. versus UK versus European Union) for progress metrics was varied. U.S. shipbuilders relied strongly on EVM-type metrics (likely because they were mandatory for most defense contracts). Although not direct measures of schedule, the U.S. shipbuilders felt that EVM metrics were good measures of progress—the exception being test and trials, where compartment/zone completions were felt to be as useful. UK shipbuilders relied less heavily than the U.S. shipbuilders on EVM-type progress metrics—focusing more

Table 8.3
Shipbuilder Use of Progress Metrics at Various Shipbuilding Phases

Fraction of Shipyards Using the Metric	Detailed Design	Construction	Test and Trials
2/3 or more	Earned value	Earned value	
2/3 to 1/3	Milestone Task Actual versus planned	Milestone Task Actual versus planned	Earned value Milestone Task
1/3 or less	Area/zone Miscellaneous	Area/zone Miscellaneous	Actual versus planned Area/zone Miscellaneous

SOURCE: Arena et al., 2005a.

on installation quantities and ratios (shown as actual versus planned in Table 8.3). However, the UK builders that did use EVM noted that it was useful in working with high-level activity schedules to measure progress. They felt that maintaining and keeping the logical dependencies for very detailed network schedules were too difficult to manage effectively. The European shipbuilders tended to use zone or area metrics and installation quantities more commonly. This usage could be a result of their higher proportion of commercial contracts where there is less change or growth on the contract and the products tend to be more MOTS-like.

To succeed in monitoring progress, EVM needs to track small, discrete levels of activity, such as work that can be finished within a week's reporting period. Tracking small-level tasks as part of measuring overall progress can help eliminate subjective judgments. The metric is straightforward: A task is either done or not. For activities that are difficult to track, such as engineering, fixed guidelines need to be established to assess progress and work scope. One U.S. shipbuilder assesses engineering progress based on a "drawing" that shows when specific content is complete. Progress is assigned when the drawing meets all the criteria for content for one of three levels. Another effective practice is to have major subcontractors report their progress and incorporate those data into overall program progress reports.

Last, the shipbuilders generally viewed that change orders/late definitions were big drivers of schedule slip. Beyond trying to minimize late changes, owners should monitor the value of unresolved changes. Using a metric for the estimated value of such changes that are not included with the baseline would help to serve as a check on the status of completeness and whether the amount of potential new work could make the schedule slip.

Oversight at Construction Shipyards

One important government responsibility entails providing oversight to the construction process at the shipyards. A cadre of government and RAN personnel should be at the shipyards to monitor progress and to resolve technical or contractual issues that may arise during the construction program. This group should provide on-site construction oversight for deviations from the design, assure compliance to quality and testing procedures, and keep the government and RAN aware of the challenges that the program faces. The on-site representatives should be experienced in both the technical and managerial aspects of delivering a major ship program and also have decisionmaking capability to facilitate concessions and deviations that have only a minor effect on cost, schedule, and performance.

In the United States, this role is played by the Supervisor of Shipbuilding (SUPSHIP), which is represented at each major U.S. naval shipbuilder. Its roles include:

- Serves as a government contract office performing administrative services for all contracts issued to the shipyard. This includes enforcing contract requirements to ensure that all parties satisfy their contractual obligations. The oversight organization may have the authority to approve minor changes to the basic contract.
- Interacts with the shipbuilder and subcontractors to improve quality and economy during the construction program.
- Monitors the progress of the build program and ensures that technical specifications on the construction of the ship are followed.
- Provides technical authority resolving and coordinating any technical issues that arise.
- Coordinates the exchange of information and correspondence between the contractor and the government.
- Performs other duties to ensure that the project is conducted and managed according to the assigned roles and responsibilities of the government and the contractors.

This is an important function that should not be overlooked. The UK *Astute* submarine program initially adopted an "eyes on, hands off" policy with the prime contractor based on the philosophy that the prime contractor was responsible for its actions and was able to deliver a submarine designed and built to the requirements and specifications. The government on-site presence was reduced from approximately 50 people to four. Furthermore, there was very little involvement from the experienced operators and maintainers of the Royal Navy. The lack of a sufficient on-site presence blinded the *Astute* program to design and construction problems that emerged in its early years.

Build-Related Factors the Program Can Control

As with the design program, a number of factors and decisions are under the control of the program. Where some design-related factors change for a pure or evolved MOTS option versus a new design option, the majority of the build-related issues hold for all three acquisition choices, assuming that at least portions of the ship will be built in Australia.

Avoid Concurrency in Design and Production

Concurrency in design and construction is an issue when construction begins before the detailed design process is sufficiently complete. It is often better and more cost-efficient to delay construction rather than risk the rework and changes that result from design immaturity. Best practices suggest that 80 percent or more of the detailed design drawings should be complete when construction begins. This was not the case in the *Collins* program, where construction of the first ship commenced when only 10 percent of the detailed drawings were complete.[14] This resulted in a significant amount of construction rework as the designs matured and were finalized.

Likewise, the AWD program has suffered from concurrency issues, with a large number of design revisions occurring after Second Pass approval and more than one-third of the way through the block construction period for the first ship in the class. As of March 2013, there had been an average of 2.75 revisions per drawing.[15] Cited causes of design revisions have been drawing errors or omissions, modifications for CFE, and government-required changes.

For a pure MOTS acquisition, it is to be expected that all, or nearly all, of the detailed design will be complete at the initiation of

---

[14] Schank et al., 2011a, p. 50.

[15] Australian National Audit Office, 2014, p. 26.

construction. However, an evolved MOTS purchase may require significant new design work. For an evolved MOTS or a new design, the Commonwealth should adopt the 80 percent rule of thumb for design completion before initiation of construction on the first platform.

Develop Effective Coordination and Quality-Control Processes to Support a Distributed Build Strategy

Shared or distributed build strategies are increasingly being adopted by shipbuilding programs. The motivation for these strategies comes mainly from the government's desire to maintain a competitive industrial base (LCS and *Virginia*), to overcome capacity limitations among shipyards (QEC), to access specific skills (DDG-1000 composite deckhouse), or to distribute work to reduce costs (Type 45).[16] In Australia, the AWD is currently being built across four shipyards: three in Australia and one in Ferrol, Spain (the parent navy shipyard). A distributed build environment can lead to increased production costs related to, for example,

- out-of-schedule completion of blocks or components by one or more shipyards
- transportation of blocks or modules between shipyards
- integration issues between blocks.

To mitigate integration issues in a distributed build environment, it is important that there be a timely exchange of data between shipyards and that effective coordination mechanisms are in place for quality control and configuration management. This may require common design software or other compatible software tools and shared networks or common data storage centers between shipyards. The DDG-1000 program suffered from schedule delays as a result of difficulties with the IT infrastructure and in sharing and operating the design software.[17] Shared IT infrastructure also may require the requisite security protocols for access to data. This may create issues when coordinating

[16] Smallman et al., 2011.

[17] GAO, 2008a, p. 10.

with overseas shipyards or could create concerns about sharing proprietary data between competing shipyards.

Ensure That Sufficient Oversight Exists at the Construction Shipyards

It is important that the program and the government be aware of the status of the build program and of any problems that exist. At the beginning of the *Astute* program, the MOD oversight at the Barrow shipyard was greatly reduced as part of the movement to control government spending. This lack of on-site presence blinded the MOD to design and construction problems that were emerging during the early years of the program. The program should have a strong presence at the shipyard to provide on-site construction oversight for deviations from design, assure compliance to quality and testing procedures, and keep the DMO aware of the challenges that the program faces. The on-site DMO representatives should be experienced in both the technical and managerial aspects of delivering a ship program and also have some decisionmaking capability to facilitate concessions and deviations that have only a minor effect on cost, schedule, or performance.

Develop a Management System to Track Progress During the Design and Build Process[18]

As with the previous lesson on sufficient oversight of the build process, the DMO must have in place a system to track progress during the build program. During the first several years of the program, there was no effective system to monitor the progress of the design and build. Ultimately, the EVM system was put in place. However, the use of the system represented a cultural change for the shipyard, and workers still find it difficult at times to allocate the proper data to the right project or task. An accurate cost-accounting system is a necessary prerequisite for a meaningful EVM system.

Earned value metrics compare the budgeted cost of work performed (BCWP) with the budgeted cost of work scheduled (BCWS)

---

[18] See Arena et al., 2005b, for a discussion of various methods used to monitor the progress of shipbuilding programs, including a specific description of EVM.

at a given point in time. When a BCWP value is less than that of the BCWS, the project is considered behind schedule. If the BCWP value exceeds the BCWS value, the project is considered ahead of schedule. The schedule performance index is equal to BCWP divided by BCWS. The cost performance index is the BCWP divided by the actual cost of work performed. An index number less than 1 indicates that the project is behind schedule and over budget.

EVM has a number of underlying assumptions and limitations. It provides few, or even incorrect, insights if the proper data are not collected and reported correctly. EVM also lacks flow and value-generation concepts. *Flow* refers to how resources and activities are sequentially related. *Value-generation work* is work performed in one time period that will allow future work to begin. Because building to sequence is so critical in shipbuilding programs, EVM must be used with care to avoid introducing poor behaviors. In addition to having an EVM system in place, it is important that it be implemented appropriately. An audit of the EVM system for the F-35 Joint Strike Fighter found it to be out of compliance, as the contractor's processes did not meet 19 of 32 DoD-required EVM guidelines.[19] Whether EVM or another progress-monitoring metric is used, it is important to have an effective system to track progress and predict cost and schedule status.

Plan for Operational Testing

Prototyping and testing of new technologies is an important step in the development of a new or modified ship class. For shipbuilding programs, prototyping of a complete system including hull form is unrealistic. Given the size and complexity, the first ship of the class may serve as the de-facto prototype. However, subsystem prototyping and testing in an appropriate shore-based test facility can reduce technical risk and is recommended when at all possible. In the *Ohio* program, use of a shore-based test facility for the combat system allowed for discovery and correction of technical problems before it was installed in

---

[19] GAO, *F-35 Joint Strike Fighter: Current Outlook Is Improved, but Long-Term Affordability is a Major Concern,* Washington, D.C.: U.S. Government Printing Office, March 2013b.

the vessel.[20] Likewise, the *Virginia* program operated on a "fly before buy" strategy. This required that new technologies be tested on land or sea and, when possible, be subjected to the entire mission profile before being incorporated into the design.[21]

It is also important to complete all or most testing before issuing production contracts. In the LCS program, the U.S. Navy has committed to the purchase of significant numbers of seaframes before the operational testing activities are complete for both the seaframes and mission modules.[22] This puts the program at risk of discovering performance issues *after* production contracts are in place and may lead to costly renegotiations or rework.

Key Points for the Manufacturing Phase

- *Avoid concurrency in design and production.* A general rule of thumb is to have a minimum of 80 percent of the detailed design complete before initiating construction of the lead ship.
- *Develop effective coordination and quality control processes to support a distributed build strategy.* Infrastructure and processes for timely exchange of data are needed in a distributed build environment.
- *Ensure that sufficient oversight exists at the construction shipyards.* The government should have an on-site presence at the shipyard to monitor and manage challenges as they arise.
- *Develop a management system to track progress during the design and build process.* This system should include appropriate metrics to measure shipbuilder performance.
- *Plan for operational testing.* Planning should include facility needs for prototype testing in a realistic environment.

---

[20] Schank et al., 2011b, pp. 24–25.

[21] Schank et al., 2011b, p. 65.

[22] GAO, *Significant Investments in the Littoral Combat Ship Continue Amid Substantial Unknowns About Capabilities, Use, and Cost,* Washington, D.C.: U.S. Government Printing Office, July 2013c, p. 6.

CHAPTER NINE

Test and Trials[1]

The last acquisition step is the test and trials phase that validates that the ship performs as expected. Some of the testing begins during the construction phase (such as material receipt and installation check), but the most significant effort begins during the trials period where total ship function is checked and any deficiencies are corrected. There is no single, correct way to go through this process. The key is that it takes significant involvement from both the builder/prime and the owner. For example, the owner has to witness and approve the results of many of the key test and trial events. It is important for the government to define the requirements for total ship testing and the supporting test plan. The contractor develops the ship subsystem test plan based on the government requirements.

Definition and Activities

Before the U.S. Navy accepts a ship, it requires that the contractor conduct a series of tests and operating and performance trials. Pre-sea trial requirements, such as dock trials, fast cruise, pre-trial audit, and combat system trial rehearsal, are required by the specifications for some ship types, particularly submarines and nuclear-powered surface ships.

[1] This section from NAVSEA, "Chapter 10: Production Acceptance Testing During Construction, Conversion and Modernization," *SUPSHIP Operations Manual,* Rev. 2, November 19, 2013.

Definition of test and trial stages[2]

Stage 1 – Material Receipt Inspection and Shop Tests: Includes those tests and inspections that provide for inventory management and physical inspection of new material, equipment and systems, and associated documentation. These tests and inspections are intended to ensure receipt of equipment in good physical condition by the shipbuilder or other industrial organization. Stage 1 documentation is not normally in the form of a test procedure. Stage 1 further includes those tests and inspections conducted prior to shipboard installation for new or repaired equipment or systems. In instances where equipment and systems are repaired aboard ship, shop test procedures may be used to validate readiness for shipboard testing. For work planning and cost accounting purposes, Stage 1 is not part of the test program and will normally be a part of the industrial organization's quality assurance program.

Stage 2 – Shipboard Installation Inspections and Tests: These are conducted prior to operation of installed or relocated equipment, cabling, waveguide, piping, ventilation, etc., to ensure that each installation has been accomplished in accordance with established plans and specifications. The shipbuilder or industrial organization is normally responsible for preparation of Stage 2 test procedures.

Stage 3 – Equipment Tests: Demonstrate that after shipboard installation, the individual equipment performs within established limits and tolerances. These equipment operability tests are conducted independent of the system (i.e., the equipment may be isolated from the system) and can be conducted prior to complete system installation.

Stage 4 – Intra-system Tests: Demonstrate that equipment and required functions, entirely within one independent system, perform within established limits and tolerances. Stage 4 testing normally consists of intra-system functions, signals, and commands within a single independent system of the combat system or ship system. Stage 4 includes all tests involving two or more items of equipment that do not involve more than one independent system of the combat system or ship system. Stage 4 tests may include tests between two or more items

[2] From NAVSEA, 2013.

of equipment and between two groups of equipment within the same "stand alone" system.

Stage 5 – Intersystem Tests: Involve testing the interfaces and interoperability between two or more independent systems within a combat system, ship system, or between the combat system and ship system. These tests demonstrate that two or more independent systems perform a specific function or functions within established standards. The exchange of intersystem signals, commands, functions, and all associated computer interfaces are included.

Stage 6 – Special Tests: Require special simulation facilities or resources external to the immediate test organization, but are conducted as part of the dockside work package for the industrial effort. Special tests can apply to one or more items of equipment, a single system, or a number of systems, and may require total ship operability. Stage 6 tests that can only be performed at-sea should be designated as Stage 7. Normally, there will be very few Stage 6 tests in an industrial test program.

Stage 7 – Trials Tests: Must be conducted during sea trials (e.g., Builder's Trials (BT), Acceptance Trials (AT), Underway Trials (UT), Combined Trials (CT), Super Trials (ST), Post-Repair Trials (PRT), and Final Contract Trials [FCT]). Test procedures are not identified with a Stage 7 number unless the test can only be conducted entirely or partially at sea.

Figure 9.1 shows notional trials milestones for a typical new construction and illustrates the sequence of testing, trials, and related events. The figure timeline begins in the upper left with construction award. The testing activities begin somewhere mid-construction where the receipt, installation, and equipment tests take place. A series of intra- and inter-system tests (Stages 4 and 5) follow the launch for the electronics, combat system, electrical generation, and main engine systems. Then, a series of trials begin where the ship is put to sea and full system performance testing begins. After the deficiencies are corrected after the acceptance trial, the ship is delivered. The Navy has ownership and performs the final fitting-out and last trial. After the final contract trial, the ship typically enters a short period (PSA) during which its systems can be updated.

Figure 9.1
Notional Major Milestones of Test and Trials During Construction

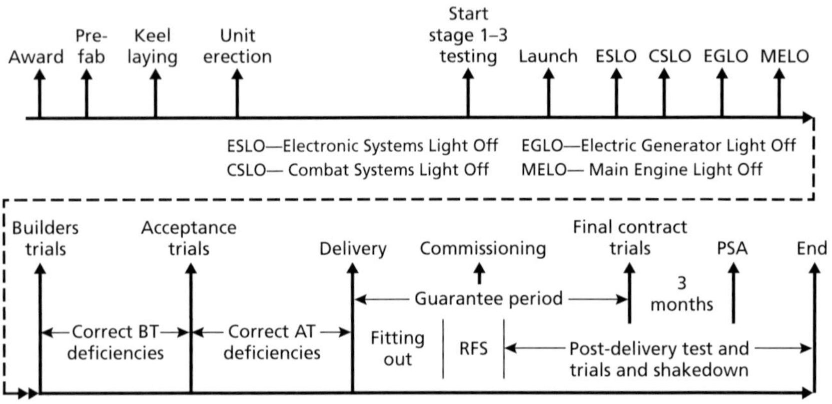

SOURCE: Adapted from NAVSEA, 2013.
NOTES: RFS = ready for sea; *light off* refers to activating the system for the first time.
RAND *RR767-9.1*

Important Considerations

Post-Trial Availability

The first availability after trials—the PSA—for the ship is an important activity with respect to design and construction planning. Because C4I systems evolve very rapidly (technical refresh can be on the order of two years or less), it is necessary to freeze these systems at some point during the design so that the design can move forward and is not continually changing. However, this means that these systems might be three or four cycles out of date when the ship is delivered, given the long design build durations for ships. The PSA updates these systems to the current technology.

Test Planning

One of the best practices that the U.S. acquisition system has identified is to develop a test and evaluation (T&E) plan early in the program (beginning in the solutions analysis phase) and maintaining and expanding it during program execution. One key document produced is called the Test and Evaluation Master Plan (TEMP). The TEMP "describes the acquisition program's planned T&E over the program's life cycle and identifies evaluation criteria for the testers. It serves as an executive summary of the overall test program. . . . The TEMP is used by the program office to

- provide an overall test management plan within the acquisition strategy bounds
- identify overall T&E activities by the government and system contractor
- guide the development of specific test events and integration of detailed test plans for those activities by summarizing relevant performance requirements
- document T&E schedule and resource requirements.[3]

The Australian acquisition process also recognizes the importance of early test planning. There are two key documents produced. At First Pass, the Test Concept Document (TCD) is produced. At Second Pass, the TCD is further refined into the Early Test Plan (ETP) and a TEMP is also produced.[4]

---

3 Defense Acquisition University, *ACQuipedia,* "Test and Evaluation Master Plan (TEMP)," July 28, 2005.

4 Australian Department of Defence, *Defence Capability Development Handbook 2012*, 2012.

CHAPTER TEN

Operations and Support[1]

In shipbuilding programs, maintenance and logistics support can occur more than a decade after the initial concept development. However, early planning for ILS should be an integral part of the requirements development, design, and construction planning process. In addition, full consideration needs to be paid in the early phases of the program to the timelines and costs for developing facilities, maintenance and training contracts, and management procedures. In this chapter we review some of the best practices for planning for and establishing an ILS program.

Begin Integrated Logistics Support Planning Early in the Program

A robust ILS plan depends on a clear concept for operations and maintenance of the vessel. Establishing and supporting a strategic plan for ILS during the design phase of the program can help to ensure that the platform can be efficiently and effectively supported through its planned life cycle. Many of Australia's problems with the operational ability of the *Collins* class resulted from the absence of a thorough ILS plan during the design and construction phase.[2] It is important for the program office to involve the technical community and the operators

---

1 This material was largely drawn from Schank et al., 2011d.

2 Schank et al., 2011a, p. 51.

in the development of an ILS plan during the design phase to ensure that the platform can be adequately supported as designed.

Equipment and materiel decisions during the requirements and design stage can have an effect on repair, replacement, and disposal costs. For example, in both the *Seawolf* and the DDG-1000 program, expensive materials were used in the platform design to reduce weight. The support and maintenance costs for these materials were underestimated, leading to cost growth and, ultimately, to decisions to reduce the number of platforms that would be procured.[3]

Maintain Adequate Funding to Develop and Execute the ILS Plan

Operational and support costs account for the vast majority of total ownership costs once a platform enters service. It is important that there be adequate funding to develop and execute the strategic ILS plan and that this funding be "protected" during the design and build phases of the program. Often, ILS funding is overlooked or even reduced in the early stages of a program, as the focus is on trying to address design or procurement issues and the resulting cost growth. For example, in the UK's *Astute* program, early cost and schedule issues led to contract renegotiations to establish new cost and schedule baselines for production; in the process, contractor logistics support decisions were deleted with an agreement that these costs would be estimated at a later date.[4]

Life-cycle support costs are difficult to accurately estimate before a ship enters service, and estimates based on previous classes of ships may not always provide reliable indicators. This is particularly true for an entirely new hull design. However, even if a MOTS or evolved MOTS platform is acquired, life-cycle costs will vary depending on the

---

[3] Schank et al., 2011b, p. 45.

[4] John F. Schank, Frank W. Lacroix, Robert E. Murphy, Cesse Ip, Mark V. Arena, and Gordon, T. Lee, *Learning from Experience, Volume III: Lessons from the United Kingdom's Astute Submarine Program,* Santa Monica, Calif.: RAND Corporation, MG-1128/3-NAVY, 2011c, p. 37.

operating environment, operational tempo, manning requirements, and other factors specific to the host Navy. For example, for the *Collins* submarine, the RAN used an existing hull design. However, the adaptations required for the RAN's operating environment also required different assumptions about life-cycle costs.[5] In addition, the level of system complexity, new technologies, or new commercial models can all affect the reliability of life-cycle cost models. Life-cycle cost estimates should use the best available knowledge and be based on the operational profile of the host Navy.

Account for Maintenance and Modernization

Maintenance of the ship and ship's systems is a critical element of readiness and sustainability. An appropriately aligned maintenance plan can help to optimize life-cycle costs for the ship class. Maintenance and modernization planning includes the facilities, processes, and schedules for both shipyard-level and shipboard maintenance.

Considering maintenance requirements during the design phase is important to ensure that there is adequate space to conduct shipboard repairs. For example, equipment requiring frequent preventive maintenance needs to be accessible by maintainers. Also, designers will have to account for large or heavy equipment that may need to be removed and repaired off site.

Developing maintenance schedules as part of the strategic plan should take into consideration the operational concept, recognizing that there needs to be time for preventive and corrective maintenance. Planning for maintenance schedules also will need to take into account equipment reliability. An in-depth knowledge of equipment reliability may involve interacting frequently with design authorities and original equipment manufacturers and developing reliability metrics and a robust database for tracking them. If Australia pursues a MOTS or evolved MOTS acquisition strategy for the Future Frigate program, there should be a solid foundation of data on reliability to support

---

[5] Schank et al., 2011a.

maintenance planning. However, data should continue to be collected through test and evaluation phases, as different operating environments and concepts may have different effects on the reliability and maintenance needs of the platform. When possible, it is important to purchase the technical data packages from all vendors. In the DDG-1000 program, failure to initially secure technical data packages led to an inability to develop a complete maintenance plan and required that the program office renegotiate with vendors for the purchase of these packages.

It is important to establish a planning-yard function to track maintenance and establish future workloads to ensure that the right maintenance is done at the right times. In particular, it is an important function of the planning yard to monitor the maintenance history of ships in the class. In addition, the planning yard should maintain contact with the design authorities and the original equipment manufacturers to keep abreast of changes to equipment maintenance requirements or procedures.

Account for Crew Training and Transition

A comprehensive ILS plan should also account for how, when, and where the crew training activities will occur. Considerations for the development of a training system should include the training concepts and strategies, simulators and training software needs, and training facilities. In addition, processes should be put into place for developing and validating operational and maintenance procedures, instructions, and manuals during the production, test, and evaluation phases.

If Australia pursues a MOTS or evolved MOTS option for the SEA 5000 program, it may be able acquire training materials and manuals under the contract. In terms of such capital assets as training facilities or simulators, Australia may eventually want to develop these domestically. However, decisions will need to be made about whether the existing RAN facilities can support initial training or if initial training should be conducted at the parent navy facilities.

It is important also to consider how crews will transition from existing platforms, and it is important that crews have had enough time with the ship and in the classroom to be familiar with the operating procedures. In the *Collins* program, the quality and timing of the training was affected by production delays, and inaccurate assumptions were made about depth of maintenance training required. This left the initial crew underprepared at the time of the first vessel's operational availability.[6]

Consider ILS from a Navy-Wide Rather Than a Program Perspective

ILS should be considered at the service level rather than at a specific program level. The current RAN fleet already has a training and maintenance infrastructure in place. The ILS program for the Future Frigate should, as much as possible, draw on any excess capacity or build on the existing infrastructure instead of developing new processes or facilities. In addition, when developing the ILS plan, the program office needs to be aware of the demands that will be placed on other facilities, especially such limited maintenance facilities as dry docks.

Although there may be some parts or systems unique to the class, all efforts should be taken to standardize parts across the fleet. This will not only improve the efficiency of supply chains but will support the health of the vendor base and potentially prevent future obsolescence issues. In the DDG-1000 program, the U.S. Navy implemented a "no unique support tools" requirement in their vendor contracts. This required that the vendors design components that could be maintained only by tools that were already in use by the Navy and prevented vendors from charging additional fees for specialized repair tools.

---

[6] Schank et al., 2011a, pp. 29–30.

Plan for Technology Advancements and Obsolescence Management

The long lead times for ship development can often lead to technology obsolescence issues.[7] Once the service life of the vessel extends beyond the technology life-cycle as defined in the initial design, the availability of parts and suppliers for older technologies may be reduced. Diminishing manufacturing sources and material shortages (DMSMS) can endanger the platform's development, production, or post-production support leading to increased costs for the Navy.

One consideration for the use of a pure MOTS acquisition strategy for Future Frigate is that any existing ship in full production will have been designed for technology that is, at a minimum, ten years old. There are two major concerns with this:

1. Is there enough available space to incorporate technology upgrades including the associated power and cooling requirements, and will incorporation of new equipment affect the structural integrity of the vessel? For example, additional equipment can add enough weight to change the ship's center of gravity, affecting ship safety and performance and contributing to redesign costs over the ship's life-cycle.[8]
2. Are there existing contracts or arrangements in place to provide support for spares and maintenance for legacy technology over the planned life cycle of the vessel?

For any of the acquisition strategies, cost savings can be achieved if a DMSMS management process is incorporated into the program early on in the design-build process. For example, the *Virginia* program established a DMSMS budget, formed a "technology refresh" Integrated Project Team, formalized standard operating procedures, and developed a memorandum of agreement with the Navy Supply Systems

---

[7] Drezner et al., 2011.

[8] See the discussion of the AMDR radar upgrades to the Flight III *Arleigh Burke* destroyers in GAO, 2012a, p. 38.

Command for advance procurement of spares. More than $124 million in cost avoidance was attributed to these initiatives.[9]

In the design phase, technology and component selection can help to prevent or mitigate obsolescence issues. Although technology needs to reach certain maturity levels before incorporating it into the ship design, it is also important that the design not include technologies that are near the functional end of their lifespans. Modular or open architecture designs are useful approaches to provide flexibility for future changes. Modular components allow for removal and upgrade or removal and replacement to accommodate new technology advances.

Key Points in Operations and Support

- *Integrated logistics support planning should begin early in the program.* ILS considerations should be incorporated into design and build decisions.
- *Maintain adequate funding to develop and execute the ILS plan.* Life-cycle costs should be estimated using the best available knowledge, and funding should be set aside from funding for design and build of the platform.
- *Consider ILS from a Navy-wide rather than a program perspective.* It is important to maximize equipment commonality and standardization across the fleet and across subsequent ships in the class.
- *Account for maintenance and modernization in ILS planning.* Consider design features that facilitate the insertion of new technologies.
- *Account for crew training and transition in ILS planning.* Planning for a training system should include the training concepts and

[9] Defense Standardization Program Office, *Diminishing Manufacturing Sources and Material Shortages; A Guidebook of Best Practices for Implementing a Robust DMSMS Management Program,* Washington, D.C.: U.S. Department of Defense, August 2012, p. 4.

strategies, simulators and training software needs, and training facilities.

- *Develop a DMSMS management process early in the design-build process to handle technology obsolescence issues.*

CHAPTER ELEVEN

Summary Comments

Naval ship acquisitions are more complex and challenging than most other weapon systems. Ship programs can take decades to design and build. Their manufacturing process is far from the production-line manufacturing process seen in most industries, which means that their acquisition process must be tailored to the ship design and building flow. They typically do not undergo full system prototyping and testing, which would be prohibitively expensive, so their design process must start with less system engineering certainty than other weapon systems. Hulls can last upward of 30 years in active service, so upgradability and sustainability must be considered during the initial design. And both the density of their outfit and their expected levels of survivability are much greater than with commercial ships, making naval combatants far more challenging.

This all points to the fact that the capability to design and build naval ships is very difficult to maintain without steady use. One significant challenge facing the SEA 5000 is an uncertain state of the domestic industrial base once the program formally begins.

The SEA 5000 program is considering three broad categories of solutions: new design, evolved MOTS, and pure MOTS. Each option entails differing risks and implications for the acquisition process and strategy. The pure MOTS solution (built outside Australia) likely would entail the least design and cost risk—as there is likely to be an experienced builder and warm supplier base. Evolved MOTS options would entail more design and build risk that would increase as the ship's design diverges from the baseline design. A new design would present the most acquisition risk as everything must start from a clean sheet. However, the operational risks would be reversed to some degree. Assuming that each option performs to specification, the clean-design specifications could be tailored to the specific needs of the RAN. The pure MOTS option is a fixed design, so the RAN would have to adapt its operations to the ship and may not get every feature that it desires.

Table 11.1 summarizes some key acquisition differences among the three options that can be observed in each of four main acquisition phases that major weapon systems go through (solutions analysis, design, construction, and operations and support); it also shows differences with respect to certain issues that cut across those phases.

In the previous chapters we discussed lessons learned as they apply to different phases of a shipbuilding program and have attempted to highlight which lessons are most applicable to the acquisition strategy selected by the RAN for the design and build of the Future Frigate. Many of these lessons will apply regardless of the selected strategy. In addition, some overarching lessons can be applied across all phases of the program, from initial requirements development to life-cycle support. In this chapter, we summarize the key overarching lessons to help guide program managers' planning and decisionmaking.

Be an Intelligent and Informed Partner in the Acquisition Process

As with any buyer of a major item, the DMO and the RAN must have knowledge and expertise in what capabilities are desired from the new ship, the current and future technologies that can affect the ship's performance, and the costs, schedules, and risks of adopting different acquisition strategies. Being an intelligent and informed member of the government-industry acquisition team requires experienced operational, technical, and managerial people. The following will help ensure the government team is fully prepared to understand the effects of various options.

Table 11.1
Key Differences Between Pure MOTS, Evolved MOTS, and New Design Acquisition Options

Pure MOTS Option	Evolved MOTS Option	New Design Option
	Solutions Analysis Phase[a]	
Choice between specific designs	Choice between specific design and level of modification	Choice between level of performance
	Design Phase (Concept, Preliminary, Contract, and Detailed)	
Requirements are selected by parent navy, not RAN	Clear requirements definition important to define level of modifications needed	Clear requirements definition critical
Can begin design process at contract design stage	Will need to progress through all design phases	Will need to progress through all design phases
Detailed design largely complete	Design periods may be short if level of modification is minimal	Greatest design risk (cost and schedule)
Construction design and instructions depend on build strategy (domestic or foreign)	Moderate design risk (cost and schedule)	Margins can be flexible
Minimal design risk (cost and schedule)	Margins are predetermined and not adjustable	
Margins are predetermined and not adjustable		
	Construction Phase	
Foreign build can leverage existing manufacturing efficiencies	Foreign build can leverage existing manufacturing efficiencies	Will start from no prior experience
	Operations and Support Phase	
Issues on IP rights and ability to modify design	Issues on IP rights and ability to modify design	Need to design for future updates and modifications
Can potentially leverage an existing parts supply base	Can potentially leverage an existing parts supply base	Least risk that the design will not satisfy RAN's needs (if requirements defined early)
Greatest risk that design will not satisfy RAN's needs	Moderate risk that design will not satisfy RAN's needs	

Table 11.1—Continued

Pure MOTS Option	Evolved MOTS Option	New Design Option
	Cross-Cutting Issues	
Acquisition strategy will be limited to a single designer	Acquisition strategy will be limited to a single designer	Many acquisition strategy options possible
Alignment between designer and builder challenging if domestic build option selected	Alignment between designer and builder challenging if domestic build option selected	Classification society open choice
Already chosen classification society (or pay for redesign)	Already chosen classification society (or pay for redesign)	Least IP challenges
Greatest IP challenges	Moderate IP challenges	

[a] Multiple options could be considered during the solutions analysis phase. However at the end of this phase, a single option is taken forward, typically.

Involve the Appropriate Organizations Early and Often

Shipbuilding programs run into problems when invalid assumptions are made or decisions are made without an adequate knowledge base or evidence to support the decisions. To develop that knowledge base, it is important to involve technical experts, industry, operators, and maintainers in each phase of the program. In the requirements phase, this varied expertise can help program managers understand both the technical and manufacturing feasibility of achieving speed, weight, or other performance capabilities. Including these experts can also help to build a better understanding of the cost trade-offs for certain capabilities or different acquisition strategies. In the design phase (for an evolved MOTS or new design), involving maintainers and operators can help program managers understand how design decisions will affect the ease of maintaining, upgrading, or replacing equipment over the life-cycle of the ship.

Clearly Assign Roles, Responsibilities, and Risks

Shipbuilding programs are more successful when there is a clear understanding of the roles, responsibilities, and risk-sharing between government and industry. These roles and responsibilities should be clearly defined in the contracting phase to prevent future disputes. The government and the contractor should be responsible for cost and schedule risk in the areas under their respective control. In addition, IP rights and ownership of technical data should be negotiated and assigned early in the program to prevent issues during design and also to support future maintenance and modernization. A clear delineation of roles and effective communication mechanisms can help to build positive and effective working relationships between government and industry.

Understand the Cost and Schedule Implications of Options

Realistic cost and schedule estimates are needed throughout the program. Cost estimates must be based on through-life costs for the fleet of ships and include cost elements ranging from design and development to operations and support to the deactivation and disposal of the ships. Well-informed cost and schedule estimates are especially important when a program begins and decisions are faced on which acquisition option provides the best value for money. The estimates during the early stages of a program may be difficult to develop if data are limited (such as with a new design) or may be readily available from responses to requests to international shipbuilders (as with a MOTS or evolved MOTS option). These early estimates will be wrong but they will be remembered and used as a benchmark throughout the program. Therefore, it is imperative that the assumptions and data that underlie the estimates be specified and that estimates be updated as additional data become available and as specific decisions are made.

Clearly State Requirements

The DMO and RAN must determine the capabilities desired from the Future Frigate. Operational requirements define the ship's missions and the operational effectiveness in accomplishing those missions. Requirements also include how the ship will operate (i.e., a CONOPS or Operational Concept Document) and be supported during its operational life. These requirements should be expressed in terms of what is desired, not how to specifically accomplish the objectives. Operational requirements are translated into such performance requirements as displacement, speed, and survivability. These operational and performance requirements are often expressed as minimum thresholds that the platform must achieve and desired objective measures whose achievement is advantageous but not critical in all circumstances. Although minimum thresholds are not commonly used in Australia's capability development process, they may be useful for evaluating MOTS or evolved MOTS options. Desired operational capabilities may be such that no existing design can meet even the minimum threshold values and a major modification of an existing design or the development of an entirely new design is the only reasonable option. Trade-offs may be needed between what is desired and what is available from existing designs.

Requirements must be clearly stated and defined in ways that are agreed to by the DMO, RAN, and the ship designer and builder. In addition to having requirements that are unambiguous and clearly stated, the government should strive to avoid any changes to requirements that may affect cost and schedule. The DMO and RAN must also define how the ship will be tested to ensure that the desired capabilities are achieved.

Understand and Obtain Required Intellectual Property Rights

When negotiating the contract for a MOTS or evolved MOTS design, it is important to obtain IP rights and ensure that there are no legal barriers for the export of documentation or materials from participating foreign suppliers. IP issues are important to address in the contract so that there are no costly delays in building, updating, or supporting the Future Frigate. These IP rights are especially important for properly modernizing and maintaining the ships during their operational lives. A clean-sheet design provides the greatest leverage in obtaining the technical data rights needed for operations, maintenance, and modernization.

Strive for Program Stability

One overarching lesson from various recent military shipbuilding programs is the importance of program stability. Stability applies in many areas: consistent funding, a long-term build strategy, fixed operational requirements, stable and capable program management, and an integrated partnership between the DMO, the RAN, and the shipbuilders. Shipbuilding programs can take more than a decade from initial concept development to full production. Over that time, there will undoubtedly be changes to the external landscape including new technological developments, industrial base developments, and shifts in national strategic or budgetary priorities. Maintaining program stability in the face of these developments requires effective mechanisms to cope with change and to manage stakeholders. It is important to account for change management processes in the contracting phase and to develop methodologies for assessing the cost and effect of changes when they do occur. The program manager also needs to manage the program's public profile through effective communication and a proactive media strategy.

Start Construction Only When Designs Are Largely Finished

Starting the construction of the lead ship before designs are close to finalized can lead to costly rework and schedule delays. In some situations, gaps in industrial base demands may favor beginning ship construction when detailed production designs are still evolving. However, it is often more cost-effective to delay construction rather than risk the rework and changes that result from design immaturity. A general rule of thumb suggests that 80 percent or more of the detailed design drawings should be complete when construction begins. With a pure MOTS design, all or nearly all of the design drawings are complete when construction begins. However, even a pure MOTS design will require changes to the product model and design drawings to accommodate the capabilities and construction practices of the shipyards. An evolved MOTS choice may require significant new design work as well as synchronization of the product model and design drawings to the construction shipyards.

Develop an Integrated Logistics Support Plan Early

A robust ILS plan depends on a clear concept of how the Future Frigate will operate and be maintained. ILS plans must be addressed early in the program and be continuously refined as the program progresses. The plan should specify what maintenance and updates are required at different points during the frigate's operational life and who will accomplish the required maintenance and modernizations. Budget pressures may lead to reductions in the development of ILS plans, since logistics support is typically viewed as less important than controlling design and construction costs. However, the majority of a ship's through-life cost occurs after the ship is delivered to the RAN. The development of ILS plans should be adequately funded and addressed during the life of the program.

Have a Strategic Perspective

The Future Frigate is only one of the RAN's strategic assets. When deciding on capabilities and requirements and selecting an acquisition strategy, it is important to consider how this platform will complement and integrate with the capabilities of other platforms. The Future Frigate program must be viewed in light of overall national objectives for the naval shipbuilding industrial base. Although it may be less expensive to build a MOTS or evolved MOTS ship in another country, national policy may dictate the use of Australian shipbuilders and suppliers. During design and build planning, it is important to consider the shipyards' other naval vessel construction workloads and timelines. This is important not only to ensure that realistic schedules can be put in place but also to coordinate distributed build efforts and to maintain industrial base skills and knowledge. It also is important to consider how the existing maintenance, training, and support infrastructure can be leveraged to reap cost savings across the fleet. When possible, program managers should try to strive for commonality in parts, tools, and materials.

Address Critical Near-Term Questions Facing the SEA 5000 Program

The SEA 5000 program needs to address at least six important, near-term questions that will shape the program for decades to come. Understanding the timing and importance of these decisions is one key to a successful program. In previous chapters, we discuss the issues facing naval ship acquisitions. Some of the key questions (and their framing issues) facing the Future Frigate program are listed below.

What Are the Operational and Performance Requirements for the Future Frigate?

- mission priorities and effectiveness
- concept of operations
- scenarios for defined missions
- operational view in the context of the broad RAN and DoD forces
- required crew levels to support operations.

Which Solution Is the Most Cost-Effective (i.e., MOTS, Evolved MOTS, New Design)?

- relative costs for each option
- relative mission effectiveness
- timing to deliver each option.

How Should Government Engage with Industry?

- mix between domestic and foreign work/content
- contractual structure/acquisition strategy
- multiple building yards versus single
- intellectual property rights
- GFE versus CFE issues.

What Are the Technical Requirements and Risks?

- new technology requirements
- testing and prototyping required before detailed design begins
- flexibility to upgrade systems cost-effectively over the life of the ship
- system integration approach to the C4I/sensors/weapons/combat system
- margins
- classification society choice
- key performance parameters.

How Will the Program Office Monitor the Program?

- independent cost and schedule assessments
- earned value management requirements

- inspection and audit activities
- testing to validate performance.

How Will the Class Be Supported Through Its Life?

- responsibilities between industry and government
- supporting infrastructure required
- refresh cycles and planning
- crew versus ashore maintenance splits.

APPENDIX A

Overview of Recent Shipbuilding Programs in Australia, the United States, and the United Kingdom

This appendix offers a brief overview of the recent shipbuilding programs that were reviewed in support of this report. Table A.1 lists the programs reviewed by country.

Table A.1
Recent Shipbuilding Programs Reviewed for This Study

Australia	United States	United Kingdom
Hobart-class AWD	Littoral Combat Ship	Type 45 *Daring*-class destroyer
Collins-class submarine	DDG-1000 *Zumwalt*-class destroyer	*Queen Elizabeth* aircraft carrier
	Virginia-class submarine	*Astute*-class submarine
	Ohio-class submarine	
	Seawolf-class submarine	

Australia Shipbuilding Programs

Air Warfare Destroyer

The SEA 4000 AWD program is the RAN's program to design, build, and deliver three *Hobart*-class guided missile destroyers. The AWD is intended to replace the RAN's *Adelaide*-class guided missile frigates, and the goals of the program are to both provide enhanced capability to the RAN's surface force and also to help sustain Australia's shipbuilding industry.[1] The AWD is based on a modified version of an existing Spanish design and is being constructed in Australia by ASC AWD Shipbuilder Pty Ltd. in a distributed-build environment. The platform incorporates the U.S. Navy's *Aegis* weapon system produced by Raytheon. The program includes a three-way alliance (henceforth "the alliance") between the DMO as the owner-participant and ASC Pty Ltd. and Raytheon as nonowner participants. The platform system design contract was awarded to Navantia S.A., a Spanish company, which elected not to be part of the alliance because of liability issues.

The AWD program entered Phase 1—the preliminary design phase—in 2002, with First Pass approval in May 2005 and Second Pass approval and commencement of the build phase in June 2007. The program has suffered from both cost growth and schedule delays. The 2013 estimates from the industry alliance suggest that the cost of the construction contract for the DDGs would be 6.8 percent in excess of the target cost estimate, and the delivery of the three DDGs has been delayed by 15 to 21 months.[2] Key issue areas have been immaturity of detailed design documentation, unstable design specifications, invalid assumptions and errors in target cost estimates, lower-than-expected shipyard productivity and subcontractor performance, and incomplete alignment of contract incentives with the platform systems gesigner.

Collins Class

In the 1970s, the RAN began planning to replace its *Oberon*-class of submarines, the first of which was slated to retire from service in the early 1990s. Australia's submarine force had been fulfilling a number of roles—maritime surveillance, maritime strike and interdiction, reconnaissance and intelligence collection, Special Forces operations, and protection of vital sea lanes—and the RAN wanted the replacement vessels, known as the *Collins* class, to be more capable in these roles than the *Oberon* fleet.

Australia intended to take an evolutionary approach in procuring the Collins class. Its initial request for tenders specified that the

---

1 Australian National Audit Office, 2014.

2 Australian National Audit Office, 2014, p. 70.

submarine employ a design already in service or would be in service by 1986. This approach was thought to mitigate the inherent risk in the country's first attempt at constructing this new class of submarines domestically. Design risks remained, however, because most conventional submarines then available that could serve as a basis for the *Collins* were designed for short-duration operations in the colder waters of the Baltic Sea.[3]

Because those submarines' operating capabilities and environments differed greatly from the *Collins*' expected performance and operating conditions, the *Collins* program ended up pursuing a developmental platform and a developmental combat system.[4] This introduced a high degree of risk into the program, which had no risk management mechanisms. Although an off-the-shelf design would not have met Australia's unique operational requirements, it would have been less risky to build.

During the *Collins* build phase, the ASC shipbuilding consortium that oversaw the program suffered business, contract, and legal problems. The main issue involved Kockums as the subcontracted designer and part owner of ASC. Once the submarines entered service, problems arose with supporting the *Collins* class, since the RAN was not properly positioned to assume the role of a parent navy. These support issues continue although recent changes have begun to result in improved operational availability of the *Collins* class.

---

[3] There were large differences in the endurance requirement and operating environments for the European and Australian submarine forces. Most European countries ran their submarines for a week at a time, departing on Monday and returning to port on Friday. Thus, their submarines were typically smaller with a lower usage rate and power requirements. The Australians, on the other hand, transited greater distances and were on station for months at a time, which had a number of implications for fuel storage, hotel services, and other hull design features. Additionally, European navies were accustomed to designing for operations in the Baltic, where the water is cold and relatively calm, which was problematic for Australia's salty, open-ocean environments and tropical waters.

[4] The *Collins* was the first class of submarines constructed in Australia. Although the RAN had experience with maintaining the *Oberon* class and had previously built commercial ships and some naval vessels, Australia's submarine construction capability had to be built from the ground up.

U.S. Shipbuilding Programs

We reviewed two recent shipbuilding programs in the United States: the LCS, and the DDG-1000 *Zumwalt* class. The concepts for these ships were developed in the 1990s and both programs are now in production. Both programs were initiated and funded in a relatively strong fiscal environment; however, shifts in national budget priorities have increased government scrutiny on program cost growth and have affected funding levels and planned quantities.

Littoral Combat Ship

The concept behind the LCS program is to provide a flexible seaframe that can accommodate various mission modules. The mission packages for the seaframe include mine countermeasures, antisubmarine warfare, and surface warfare missions. The program concept development began in 2002 and system development began in May 2004. There were several innovative concepts underpinning the program:

- the employment of modular weapon systems
- highly reduced manning levels
- heavy reliance on shore-based contractor maintenance.

In addition, the Navy pursued an original acquisition strategy. The Navy purchased two unique designs for construction in different shipyards; the Lockheed Martin *Freedom* variant—a monohull design—is being built in Marinette, Wisconsin, and the Austal USA *Independence* variant—a trimaran—is being built in Mobile, Alabama. The production decisions for the first four hulls were in December 2004 and October 2005. The lead-ship deliveries for both designs were in 2008 and 2009. Whereas the initial acquisition strategy was to initially fund both designs and then select one design for continued production, the Navy has changed this strategy and has continued to procure both seaframes.

As of 2014, the program has seen significant increases (greater than 148 percent) in development costs from the initial estimates, and per-unit costs have grown by 61.6 percent.[5] One cited issue for cost growth was the decision to enter production of LCS hulls 1 and 2 without a stable design or production processes.[6] Other program issues include integration challenges with the combat systems suite and the mission modules, poor workmanship, and inability to meet initial capability requirements. In addition, in 2013, it was determined that crew size needs were greater than originally underestimated requiring costly redesign and back-fitting the hull for additional berthing space and increased hotel services capacity. The program continues to face external pressures and a loss of political confidence, resulting in funding and quantity changes and general uncertainty about the future of the program.

DDG-1000 *Zumwalt* Class

The DDG-1000 *Zumwalt* class is a 15,000-ton multimission destroyer designed to provide advanced land-attack capability in support of forces ashore and littoral operations. System development started in 2004, with the production decision in November 2005 and construction of the lead ship initiated in February 2009. Design and construction of the hulls are shared between two shipyards: Huntington Ingalls Industries shipyard in Gulfport, Mississippi, and General Dynamics' Bath Iron Works yard in Maine. The systems development contracts for the ships are held with Raytheon and BAE Systems with the Navy responsible for integration. The contract for the lead ship is a cost-plus type contract.

As of 2014, program development costs have grown approximately 342 percent from the baseline estimate and program unit costs have grown by 541.1 percent.[7] Some of the program unit cost growth can be attributed to government decisions regarding reductions in total quantity from 32 planned hulls in 1998 to three hulls in 2010. This decision has spread the R&D costs over a fewer number of plat-

---

[5] GAO, *Defense Acquisitions; Assessments of Selected Weapon Programs,* Washington, D.C.: U.S. Government Printing Office, March 2014, p. 19.

[6] GAO, 2014, p. 94.

[7] GAO, 2014, p. 19.

forms. Total program costs have actually decreased by approximately 40 percent because of quantity changes. Other key challenges for this program have been the integration of new technologies—particularly with integration of a planned composite deckhouse—which has had required significant rework and has caused schedule delays and cost increases. Other cited issues are technical immaturity of critical technologies; only three of the 11 critical technologies will have been demonstrated in a realistic environment at the point of installation on the lead ship of the class.[8]

We also conducted a review of lessons learned from previous work that RAND undertook for three recent submarine design and build programs in the United States; the *Ohio, Seawolf,* and *Virginia*-class submarines. The *Ohio*-class ballistic missile submarine was conceived as a replacement for the aging SSBN fleet and sought to provide the fleet with a more capable and stealthy ballistic missile delivery platform. Concept development for the submarine began in the early 1970s. Electric Boat was awarded both the design and build contracts for the platform. The *Ohio* was the first submarine program to use modular-build construction. The first boat of the class was delivered in October 1981 with a 31-month delay from the original target delivery. Cost growth for Flight I of the class was 21 percent over the target cost. Some of the key issues affecting the program were unrealistic initial cost and schedule estimates and insufficient management of supplier quality and reliability for both government- and contractor-furnished equipment. External factors also affected cost and schedule; these included shipyard labor disputes and workforce shortages. One area where the *Ohio* program was successful was in minimizing technical risk. The program also benefitted from an external environment with a strong industrial base and ample funding.

Seawolf Class

The *Seawolf*-class attack submarine program was initiated in the early 1980s as a follow-on platform to the *Los Angeles* class and was undertaken in response to an evolving maritime strategy and what was seen

[8] GAO, 2014, p. 60.

as clear ASW advances by the Soviet Union. Unlike the *Ohio* program, which had taken a more pragmatic approach to technology risk, the *Seawolf* program pushed the technology boundaries of the time. The aggressive operational capabilities desired for *Seawolf* drove early decisions to select advanced and unproven or immature technologies for the design.[9] The acquisition strategy involved a split-design between Electric Boat and Northrup Grumman Newport News. The split-design strategy and high levels of technological risk created both cost and schedule issues for the program through the design and build phases. In addition, the *Seawolf* program fell victim to external influences, including shifting national maritime and budget priorities; this caused the program to be truncated from a planned class of 29 vessels to only three vessels. Design and construction of the *Seawolf* spanned 15 years from early concept development in 1982 to commissioning of the first ship in 1997.

***Virginia* Class**

The SSN 774 *Virginia*-class attack submarine is a multimission platform designed for littoral operations while still maintaining traditional open-ocean antisubmarine and antisurface capability. It was developed in the early 1990s as a successor to the *Los Angeles* and *Seawolf* classes and incorporated many of the lessons learned from the struggles of the *Seawolf* program. Design and construction spanned 16 years from the original concept development in 1988 to the delivery of the first boat in 2004. The *Virginia* was designed by Electric Boat and built by both Electric Boat and Northrup Grumman Newport News (now Huntington Ingalls Industries Newport News Shipbuilding) shipyard using a modular build process.

The *Virginia* program has generally been considered a success story with a schedule delay of only four months for delivery of the first vessel and cost overruns of only 8 percent of budgeted cost and

---

[9] For example, the use of HY-100 steel, which would allow greater depths for the submarine with reduced weight penalties. Although HY-100 had been tested to depth on sections of the *Los Angeles* class, the production and welding processes for the material were not well developed.

2 percent of target cost. Successes have been attributed to an effective contracting strategy that included appropriate performance incentives to management innovations such as the IPPD technique to effective oversight by the program office. Cost overruns were related to labor productivity issues, integration issues between shipyards, and increases in material costs resulting from a diminished vendor base and lack of design maturity for certain components.[10]

United Kingdom Shipbuilding Programs

The UK has undertaken three major shipbuilding programs in the past two decades. These are the Type 45 *Daring*-class destroyer, the *Queen Elizabeth*-class carrier, and the *Astute* submarine. In the next section, we provide a brief overview of these programs' cost and schedule overruns and some key issues that have been cited as root causes for those overruns.

Type 45 Class

The UK's Type 45, *Daring*-class destroyer program was intended to replace the Type 42 destroyer and bring an enhanced antiair warfare capability, greater efficiency, and more adaptable design.[11] Originally, the MOD planned to acquire 12 ships, but budget constraints reduced the planned number to six. The program was approved in 2000 and procured under a fixed-price contract for the first three ships of the class. The Type 45 program is currently being managed and delivered by BAE Systems Surface Ships. The program is using a shared-build process between two shipyards and has delivered the lead and second ship to the Royal Navy.

The Type 45 program initially suffered from an acquisition and contracting strategy that did not appropriately allocate risk and from difficulties in the relationships between the government and the prime contractor. By 2009, the program had experienced significant cost

---

[10] Schank et al., 2011b.

[11] National Audit Office, 2009, p. 9.

overruns of over 30 percent from the initial estimates. In 2010, a program restructuring was initiated to try to control costs. As of 2012, the total program cost was estimated to be 17 percent above the original estimate at program approval, and the estimated in-service date was delayed by 38 months from original estimate.[12]

Queen Elizabeth Class

The *Queen Elizabeth*-class program is intended to add two aircraft carriers to the UK's surface fleet with an operational military capability available by 2020. The design process was initiated in 1999, and the design concept provided for a modular construction, as no single shipyard in the UK had the capacity to solely deliver a single ship. In 2005, the MOD formed the aircraft carrier alliance (ACA), a consortium of government and industry that includes the MOD (as the customer), BAE Systems, Babcock International Group, and Thales Naval, for the management and delivery of the two carriers. The construction of the carriers, which began in 2009, is being undertaken across seven shipyards with block integration and assembly to occur at the Rosyth dockyard in Scotland.

The *Queen Elizabeth*-class program has been beset by rising cost growth and schedule delays attributable to a number of factors, including extended contract negotiations, technical issues, and design changes related to the operation of a different variant of the Joint Strike Fighter (JSF-35). As of 2013, the projected in-service date has been delayed by 29 months, and the forecasted cost to completion has grown by 72 percent from the original estimate. In November 2013, the MOD concluded renegotiations with industry to try to address cost growth through program restructuring. The deal reached with the ACA is expected to identify cost savings and shares the future cost growth equally between the MOD and industry.

---

[12] National Audit Office, *The Major Projects Report 2012,* London: The Stationery Office, December 17, 2012, p. 32.

Astute Class

In the 1980s, the UK MOD conducted a number of studies to choose a replacement for the *Swiftsure* and *Trafalgar* classes of attack submarines. Whereas *Trafalgar* was a close derivative of *Swiftsure*, the goal for the SSN20 (the original name for the new project) was a new submarine with a major upgrade in capability. The Russian threat during the Cold War was significant and the UK, much like the United States with the *Seawolf* design, was seeking to counter Soviet advances in antisubmarine warfare and in ballistic missile submarine capabilities.

The original cost estimates for the new submarine were significantly higher than those for the previous classes, reflecting the desire for a revolutionary design rather than an evolutionary design with enhanced capabilities. But as the initial feasibility studies drew to a close, the Berlin Wall fell and the Cold War ended, prompting some policymakers to question whether the enhanced and costly capabilities of the SSN20 were necessary. As a result, new studies were conducted with cost control as the main objective. These studies led to what was referred to as the Batch 2 Trafalgar and ultimately to the *Astute* class.

The design and construction of the *Astute* class faced several problems, including a significant gap between the design and build of the *Vanguard* class and the new *Astute* class, the changing role of government, which saw the MOD taking an "eyes on, hands off" approach with the new class, and several changes in the ownership of the UK shipyard that had built the vast majority of UK nuclear submarines. These problems resulted in a significant increase in cost and a lengthy schedule delay. The first three ships of the class experienced an in-service delay of 58 months and a cost overrun of approximately 50 percent from estimated cost at program approval. Many of the challenges with the *Astute* program have been overcome, although the production schedule is still at issue and may cause some disruption to the successor program that will replace the *Vanguard*-class ballistic missile submarines. The first two ships of the planned class of seven submarines had completed construction and a major portion of their sea trials by 2013.

APPENDIX B

Technology Readiness Levels

Technology readiness levels were first developed by the National Aeronautical and Space Administration and are frequently used as an indicator of technology risk in major acquisition programs. TRLs are measured on a scale of 1 to 9 as in Table B.1, beginning with basic research and, at the highest level, a fully integrated commercial product.

Table B.1
Technology Readiness Levels Definitions

TRL	Description
1	Basic principles observed and reported
2	Technology concept or application formulated
3	Analytic and experimental critical function/proof of concept
4	Component validation in laboratory environment
5	Component validation in relevant environment
6	System/subsystem prototype demonstrated in relevant environment
7	System prototype demonstrated in realistic environment
8	Actual system completed and qualified through test and evaluation
9	Actual system proven through successful mission operations

SOURCE: GAO, *Defense Acquisitions; Assessments of Selected Weapon Programs*, Washington, D.C.: U.S. Government Printing Office, March 2013a.

APPENDIX C

Best Practices in Scheduling

Table C.1
GAO Scheduling Best Practices

Capturing All Activities
The schedule should reflect all activities as defined in the project's work breakdown structure, which defines in detail the work necessary to accomplish a project's objectives, including activities both the owner and contractors are to perform.
Sequencing All Activities
The schedule should be planned so that critical project dates can be met. To do this, activities need to be logically sequenced—that is, listed in the order in which they are to be carried out. In particular, activities that must be completed before other activities can begin (predecessor activities), as well as activities that cannot begin until other activities are completed (successor activities), should be identified. Date constraints and lags should be minimized and justified. This helps ensure that the interdependence of activities that collectively lead to the completion of events or milestones can be established and used to guide work and measure progress.
Assigning Resources to All Activities
The schedule should reflect the resources (labor, materials, overhead) needed to do the work, whether they will be available when needed, and any funding or time constraints.
Establishing the Duration of All Activities
The schedule should realistically reflect how long each activity will take. When the duration of each activity is determined, the same rationale, historical data, and assumptions used for cost estimating should be used. Durations should be reasonably short and meaningful and allow for discrete progress measurement. Schedules that contain planning and summary planning packages as activities will normally reflect longer durations until broken into work packages or specific activities.

Table C.1—Continued

Verifying That the Schedule Can Be Traced Horizontally and Vertically
The detailed schedule should be horizontally traceable, meaning that it should link products and outcomes associated with other sequenced activities. These links are commonly referred to as "hand-offs" and serve to verify that activities are arranged in the right order for achieving aggregated products or outcomes. The integrated master schedule should also be vertically traceable—that is, varying levels of activities and supporting subactivities can be traced. Such mapping or alignment of levels enables different groups to work to the same master schedule.
Confirming That the Critical Path Is Valid
The schedule should identify the program critical path—the path of longest duration through the sequence of activities. Establishing a valid critical path is necessary for examining the effects of any activity's slipping along this path. The program critical path determines the program's earliest completion date and focuses the team's energy and management's attention on the activities that will lead to the project's success.
Ensuring Reasonable Total Float
The schedule should identify reasonable float (or slack)—the amount of time by which a predecessor activity can slip before the delay affects the program's estimated finish date—so that the schedule's flexibility can be determined. Large total float on an activity or path indicates that the activity or path can be delayed without jeopardizing the finish date. The length of delay that can be accommodated without the finish date's slipping depends on a variety of factors, including the number of date constraints within the schedule and the amount of uncertainty in the duration estimates, but the activity's total float provides a reasonable estimate of this value. As a general rule, activities along the critical path have the least float.
Conducting a Schedule Risk Analysis
A schedule risk analysis uses a good critical path method schedule, data about project schedule risks and opportunities, and statistical simulation to predict the level of confidence in meeting a program's completion date, determine the time contingency needed for a level of confidence, and identify high-priority risks and opportunities. As a result, the baseline schedule should include a buffer or reserve of extra time.
Updating the Schedule Using Actual Progress and Logic
Progress updates and logic provide a realistic forecast of the start and completion dates for program activities. Maintaining the integrity of the schedule logic at regular intervals is necessary to reflect the true status of the program. To ensure that the schedule is properly updated, the people responsible for the updating should be trained in critical path method scheduling.

Table C.1—Continued

Maintaining a Baseline Schedule
A baseline schedule is the basis for managing the project scope, the time period for accomplishing it, and the required resources. The baseline schedule is designated the target schedule, subject to a configuration management control process, against which project performance can be measured, monitored, and reported. The schedule should be continually monitored to reveal when forecasted completion dates differ from planned dates and whether schedule variances will affect downstream work. A corresponding baseline document explains the overall approach to the project, defines custom fields in the schedule file, details ground rules and assumptions used in developing the schedule, and justifies constraints, lags, long activity durations, and any other unique features of the schedule.

SOURCE: GAO, *GAO Schedule Assessment Guide; Best Practices for Project Schedules*, Washington, D.C.: U.S. Government Printing Office, May 2012b.

Bibliography

Arena, Mark V., John Birkler, John F. Schank, Jessie Riposo, and Clifford A. Grammich, *Monitoring the Progress of Shipbuilding Programmes: How Can the Defence Procurement Agency More Accurately Monitor Progress*? Santa Monica, Calif.: RAND Corporation, MG-235-MOD, 2005a. As of October 7, 2014: http://www.rand.org/pubs/monographs/MG235.html

Arena, Mark V., Hans Pung, Cynthia R. Cook, Jefferson P. Marquis, Jessie Riposo, and Gordon T. Lee, *The United Kingdom's Naval Shipbuilding Industrial Base: The Next Fifteen Years,* Santa Monica, Calif.: RAND Corporation, MG-294-MOD, 2005b. As of October 7, 2014: http://www.rand.org/pubs/monographs/MG294.html

Arena, Mark V., Irv Blickstein, Obaid Younossi, and Clifford A. Grammich, *Why Has the Cost of Navy Ships Risen? A Macroscopic Examination of the Trends in U.S. Naval Ship Costs over the Past Several Decades,* Santa Monica, Calif.: RAND Corporation, MG-484-NAVY, 2006. As of October 7, 2014: http://www.rand.org/pubs/monographs/MG484.html

Arena, Mark V., Irv Blickstein, Abby Doll, Jeffrey A. Drezner, James G. Kallimani, Jennifer Kavanagh, Daniel F. McCaffrey, Megan McKernan, Charles Nemfakos, Rena Rudavsky, Jerry M. Sollinger, Daniel Tremblay, and Carolyn Wong, *Management Perspectives Pertaining to Root Cause Analyses of Nunn-McCurdy Breaches; Volume 4: Program Manager Tenure, Oversight of Acquisition Category II Programs, and Framing Assumptions,* Santa Monica, Calif.: RAND Corporation, MG-1171/4-OSD, 2013. As of October 7, 2014: http://www.rand.org/pubs/monographs/MG1171z4.html

———, *Defence Capability Development Handbook 2012*, 2012.

Australian Department of Defence, Defence Materiel Organisation, *Defence Materiel Handbook (Engineering Management): Defence Capabilities Document Guide,* Version 1.0, DMH (ENG) 12-3-003, November 2011.

———, *Future Submarine Industry Skills Plan: A Plan for the Naval Shipbuilding Industry,* 2013. As of October 9, 2014: http://www.defence.gov.au/dmo/Multimedia/FSISPWEB-9-4506.pdf

Australian National Audit Office, *The Major Projects Report 2013,* London: The Stationery Office, February 13, 2013a.

———, *2012–13 Major Projects Report,* Canberra, Australia, 2013b.

———, *Air Warfare Destroyer Program; Audit Report No. 22 2013–14,* Canberra, Ausralia, 2014.

Birkler, John, John F. Schank, Jessie Riposo, Mark V. Arena, Robert W. Button, Paul DeLuca, James Dullea, James G. Kallimani, John Leadmon, Gordon T. Lee, Brian McInnis, Robert Murphy, Joel B. Predd, and Raymond H. Williams, *Australia's Submarine Design Capabilities and Capacities: Challenges and Options for the Future Submarine,* Santa Monica, Calif.: RAND Corporation, MG-1033-AUS, 2011. As of October 7, 2014:
http://www.rand.org/pubs/monographs/MG1033.html

Blickstein, Irv, Michael Boito, Jeffrey A. Drezner, James Dryden, Kenneth Horn, James G. Kallimani, Martin C. Libicki, Megan McKernan, Roger C. Molander, Charles Nemfakos, Chad J. R. Ohlandt, Caroline R. Milne, Rena Rudavsky, Jerry M. Sollinger, Katherine Watkins Webb, and Carolyn Wong, *Root Cause Analyses of Nunn-McCurdy Breaches, Volume 1: Zumwalt-Class Destroyer, Joint Strike Fighter, Longbow Apache, and Wideband Global Satellite,* Santa Monica, Calif.: RAND Corporation, MG-1171/1-OSD, 2011. As of October 7, 2014:
http://www.rand.org/pubs/monographs/MG1171z1.html

Blickstein, Irv, Jeffrey A. Drezner, Martin C. Libicki, Brian McInnis, Megan McKernan, Charles Nemfakos, Jerry M. Sollinger, and Carolyn Wong, *Root Cause Analyses of Nunn-McCurdy Breaches, Volume 2: Excalibur Artillery Projectile and the Navy Enterprise Resource Planning Program, with an Approach to Analyzing Complexity and Risk,* Santa Monica, Calif.: RAND Corporation, MG-1171/2-OSD, 2012. As of October 7, 2014:
http://www.rand.org/pubs/monographs/MG1171z2.html

Bolten, Joseph G., Robert S. Leonard, Mark V. Arena, Obaid Younossi, and Jerry M. Sollinger, *Sources of Weapon System Cost Growth: Analysis of 35 Major Defense Programs,* Santa Monica, Calif.: RAND Corporation, MG-670-AF, 2008. As of October 7, 2014:
http://www.rand.org/pubs/monographs/MG670.html

Brown, Gary, and Cliff Harris, "Matching Product Development Practices to the Product Life Cycle," Center for the Management of Technological and Organizational Change (CMTOC), *Highlights of the Thirty-Fifth Advanced Manufacturing Forum*, February 27–March 1, 1995.

Cavas, Christopher P., "LCS: 'Considerable Cost Overruns,'" *Navy Times,* January 11, 2007. As of June 29, 2010:
http://www.navytimes.com/news/2007/01/dfnLCScostsweb07011/

Davies, John Paul, *Alliance Contracts and Public Sector Governance,* Ph.D. thesis, Griffith Law School, Griffith University, South East Queensland, Australia, August 2008.

Deaton, William A., and James L. Conklin, "Developing Reconfigurable Command Spaces for the Ford Class Aircraft Carriers," *Engineering and Total Ship Symposium 2010,* American Society of Naval Engineers, June 2010.

Defense Acquisition University, *Defense Acquisition Guidebook,* January 10, 2012.

———, *ACQuipedia,* "Test and Evaluation Master Plan (TEMP)," July 28, 2005.

DefenceSA, *Naval Shipbuilding: Australia's $250 Billion Nation Building Opportunity,* undated.

Defence Materiel Organisation, *Defence Procurement Policy Manual (DPPM): Mandatory Procurement Guidance for Defence and DMO Staff,* July 1, 2013.

Defense Standardization Program Office, *Diminishing Manufacturing Sources and Material Shortages; A Guidebook of Best Practices for Implementing a Robust DMSMS Management Program,* Washington, D.C.: U.S. Department of Defense, August 2012.

DMO—*See* Defence Materiel Organisation.

Drezner, Jeffrey A., Mark V. Arena, Megan P. McKernan, Robert E. Murphy, and Jessie Riposo, *Are Ships Different? Policies and Procedures for the Acquisition of Ship Programs,* Santa Monica, Calif.: RAND Corporation, MG-991-OSD/NAVY, 2011. As of October 7, 2014:
http://www.rand.org/pubs/monographs/MG991.html

Etter, Delores, Paul E. Sullivan, Charles S. Hamilton, and Barry J. McCullough, *Statement before Subcommittee on Seapower and Expeditionary Forces of the House Armed Services Committee on Acquisition Oversight of the U.S. Navy's Littoral Combat Ship Program,* February 8, 2007.

Ewing, Philip, "20 to Join LCS Crew on Trial Deployment," *Navy Times,* November 14, 2009. As of June 29, 2010:
http://www.navytimes.com/news/2009/11/navy_freedom_deployment_111409w/

GAO—*See* U.S. Government Accountability Office.

General Dynamics Electric Boat, *The VIRGINIA Class Submarine Program: A Case Study,* Groton, Conn., February 2002.

Hess, Ronald Wayne, Denis Rushworth, Michael Hynes, and John E. Peters, *Disposal Options for Ships,* Santa Monica, Calif.: RAND Corporation, MR-1377-NAVY, 2001. As of October 7, 2014:
http://www.rand.org/pubs/monograph_reports/MR1377.html

International Association of Classification Societies, *Classification Societies—What Why and How?* June 2011.

Martin, Edward C., "Incentive Contracting," PowerPoint file, SAF/AQC Field Support Team, April 25, 2011, p. 9.

McIntosh, Malcolm K., and John B. Prescott, *Report to the Minister for Defence on the Collins Class Submarine and Related Matters,* Canberra: CanPrint Communications Pty Ltd., June 1999.

Mortimer, David, *Going to the Next Level: The Report of the Defence Procurement and Sustainment Review,* Defence Materiel Organisation, 2008.

NAO—*See* National Audit Office.

National Audit Office, *Providing Anti-Air Warfare Capability: The Type 45 Destroyer,* London: The Stationery Office, 2009.

———, Ministry of Defence, *The Major Projects Report 2011: Appendices and Project Summary Sheets,* HC 1520-I, Session 2010–2012, November 16, 2011.

———, *The Major Projects Report 2012,* London: The Stationery Office, December 17, 2012.

———, *The Major Projects Report 2013,* London: The Stationery Office, February 13, 2013.

Naval Sea Systems Command, *Description of the Naval Ship Design Phases,* briefing, Washington, D.C., February 2008.

———, "Chapter 10: Production Acceptance Testing During Construction, Conversion and Modernization," *SUPSHIP Operations Manual,* Rev. 2, November 19, 2013.

Naval Sea Systems Command, Shipbuilding Support Office, *Naval Vessel Register: Inventory of US Naval Ships and Service Craft,* NVR Online, updated October 6, 2014. As of October 7, 2014:
http://www.nvr.navy.mil/index.htm

NAVSEA—*See* Naval Sea Systems Command.

Navy League of the United States, "Shipbuilding Milestones," undated. As of October 7, 2014:
http://www.mynavyleague.org/commissionings/ShipbuildingMilestones.php

O'Rourke, Ronald, *Navy Littoral Combat Ship (LCS) Program: Background and Issues for Congress,* Washington, D.C.: Congressional Research Service, RL33741, August 4, 2014a.

———, *Navy DDG-51 and DDG-1000 Destroyer Programs: Background and Issues for Congress,* Washington, D.C.: Congressional Research Service, RL32109, April 8, 2014b.

Office of the Deputy Director of Defense Procurement and Acquisition Policy for Cost, Pricing, and Finance, *Contract Pricing Reference Guide: Volume 4,* 2012.

Office of the Under Secretary of Defense, Acquisition, Technology and Logistics, Defense Procurement and Acquisition Policy, *Department of Defense COR Handbook,* March 12, 2012.

Page, Jonathan, *Flexibility in Early Stage Design of US Navy Ships: An Analysis of Options,* master's thesis, Engineering Systems Division and the Department of Mechanical Engineering, Massachusetts Institute of Technology, Cambridge, June 2011.

Program Executive Office (PEO), Ships, U.S. Navy, "What Is the Difference Between LCS and JHSV?" undated.

Pung, Hans, Laurence Smallman, Mark V. Arena, James G. Kallimani, Gordon T. Lee, Samir Puri, and John F. Schank, *Sustaining Key Skills in the UK Naval Industry,* Santa Monica, Calif.: RAND Corporation, MG-725-MOD, 2008. As of October 7, 2014:
http://www.rand.org/pubs/monographs/MG725.html

Schank, John F., Hans Pung, Gordon T. Lee, Mark V. Arena, and John Birkler, *Outsourcing and Outfitting Practices: Implications for the Ministry of Defence Shipbuilding Programmes,* Santa Monica, Calif.: RAND Corporation, MG-198-MOD, 2005a. As of October 7, 2014:
http://www.rand.org/pubs/monographs/MG198.html

Schank, John F., Jessie Riposo, John Birkler, and James Chiesa, *The United Kingdom's Nuclear Submarine Industrial Base, Volume 1: Sustaining Design and Production Resources,* Santa Monica, Calif.: RAND Corporation, MG-326/1-MOD, 2005b. As of October 7, 2014:
http://www.rand.org/pubs/monographs/MG326z1.html

Schank, John F., Cynthia R. Cook, Robert Murphy, James Chiesa, Hans Pung, and John Birkler, *The United Kingdom's Nuclear Submarine Industrial Base, Volume 2: Ministry of Defence Roles and Required Technical Resources,* Santa Monica, Calif.: RAND Corporation, MG-326/2-MOD, 2005c. As of October 7, 2014:
http://www.rand.org/pubs/monographs/MG326z2.html

Schank, John F., Mark V. Arena, Paul DeLuca, Jessie Riposo, Kimberley Curry Hall, Todd Weeks, and James Chiesa, *Sustaining Nuclear Submarine Design Capabilities,* Santa Monica, Calif.: RAND Corporation, MG-608-NAVY, 2007. As of October 7, 2014:
http://www.rand.org/pubs/monographs/MG608.html

Schank, John F., Christopher G. Pernin, Mark V. Arena, Carter C. Price, and Susan K. Woodward, *Controlling the Cost of C4I Upgrades on Naval Ships,* Santa Monica, Calif.: RAND Corporation, MG-907-NAVY, 2009. As of October 7, 2014:
http://www.rand.org/pubs/monographs/MG907.html

Schank, John F., Cesse Cameron Ip, Kristy N. Kamarck, Robert E. Murphy, Mark V. Arena, Francis W. Lacroix, and Gordon T. Lee, *Learning from Experience, Volume IV: Lessons from Australia's Collins Submarine Program,* Santa Monica, Calif.: RAND Corporation, MG-1128/4-NAVY, 2011a. As of October 7, 2014: http://www.rand.org/pubs/monographs/MG1128z4.html

Schank, John F., Cesse Ip, Francis W. Lacroix, Robert E. Murphy, Mark V. Arena, Kristy N. Kamarck, and Gordon T. Lee, *Learning from Experience, Volume II: Lessons from the U.S. Navy's Ohio, Seawolf, and Virginia Submarine Programs,* Santa Monica, Calif.: RAND Corporation, MG-1128/2-NAVY, 2011b. As of October 8, 2014:
http://www.rand.org/pubs/monographs/MG1128z2.html

Schank, John F., Francis W. Lacroix, Robert E. Murphy, Cesse Ip, Mark V. Arena, and Gordon, T. Lee, *Learning from Experience, Volume III: Lessons from the United Kingdom's Astute Submarine Program,* Santa Monica, Calif.: RAND Corporation, MG-1128/3-NAVY, 2011c. As of October 8, 2014:
http://www.rand.org/pubs/monographs/MG1128z3.html

Schank, John F., Francis W. LaCroix, Robert E. Murphy, Mark V. Arena, and Gordon T. Lee, *Learning from Experience, Volume I: Learning from Experience: Lessons from the Submarine Programs of the United States, United Kingdom, and Australia,* Santa Monica, Calif.: RAND Corporation, MG-1128/1-NAVY, 2011d. As of October 8, 2014:
http://www.rand.org/pubs/monographs/MG1128z1.html

Schank, John F., Scott Savitz, Ken Munson, Brian Perkinson, James McGee, and Jerry Sollinger, *Designing Adaptable Ships: Modularity and Flexibility in Future Ship Designs,* Santa Monica, Calif.: RAND Corporation, RR-696-NAVY, forthcoming.

Society of Allied Weight Engineers, Inc., *Weight Estimating and Margin Manual for Marine Vehicles,* Marine Systems Government–Industry Workshop, Society of Allied Weight Engineers, Recommended Practice Number 14, May 22, 2001.

Smallman, Laurence, Hanlin Tang, John F. Schank, and Stephanie Pezard, *Shared Modular Build of Warships: How a Shared Build Can Support Future Shipbuilding,* Santa Monica, Calif.: RAND Corporation, TR-852-NAVY, 2011. As of October 8, 2014:
http://www.rand.org/pubs/technical_reports/TR852.html

Summers, Wilson, IV, "'Incentivizing'—An Effective Motivator? Grappling with Defense Contractor Incentive Issues," *Program Manager,* March–April 1995, pp. 26–28.

U.S. Department of Defence, *Earned Value Management Implementation Guide,* October 2006.

U.S. Government, "Rights in Technical Data," *Defense Federal Acquisition Regulations Supplement,* Subpart 227.71, February 28, 2014.

U.S. Government Accountability Office, *Best Practices: Capturing Design and Manufacturing Knowledge Early Improves Acquisition Outcomes,* Washington, D.C.: U.S. Government Printing Office, GAO-02-701, July 2002.

———, *Defense Acquisitions; Zumwalt-Class Destroyer Program Emblematic of Challenges Facing Navy Shipbuilding,* Washington, D.C.: U.S. Government Printing Office, July 2008a.

———, *Defense Acquisitions: Cost to Deliver Zumwalt-Class Destroyers Likely to Exceed Budget; Report to the Subcommittee on Armed Services,* U.S. Senate, Washington, D.C.: U.S. Government Printing Office, July 2008b.

———, *Best Practices: High Levels of Knowledge at Key Points Differentiate Commercial Shipbuilding from Navy Shipbuilding,* Washington, D.C.: U.S. Government Printing Office, May 2009.

———, *Littoral Combat Ship: Actions Needed to Improve Operating Cost Estimates and Mitigate Risks in Implementing New Concepts,* Washington, D.C.: U.S. Government Printing Office, February 2010.

———, *Arleigh Burke Destroyers; Additional Analysis and Oversight Required to Support the Navy's Future Surface Combatant Plans,* Washington, D.C.: U.S. Government Printing Office, January 2012a.

———, *GAO Schedule Assessment Guide; Best Practices for Project Schedules,* Washington, D.C.: U.S. Government Printing Office, May 2012b.

———, *Defense Acquisitions; Assessments of Selected Weapon Programs,* Washington, D.C.: U.S. Government Printing Office, March 2013a.

———, *F-35 Joint Strike Fighter: Current Outlook Is Improved, But Long-Term Affordability Is a Major Concern,* Washington, D.C.: U.S. Government Printing Office, March 2013b.

———, *Significant Investments in the Littoral Combat Ship Continue Amid Substantial Unknowns About Capabilities, Use, and Cost,* Washington, D.C.: U.S. Government Printing Office, July 2013c.

———, *Defense Acquisitions; Assessments of Selected Weapon Programs,* Washington, D.C.: U.S. Government Printing Office, March 2014.

U.S. House of Representatives, 111th Congress, 1st Session, *Weapons Systems Reform Act of 2009,* Public Law 111-23, Washington, D.C.: U.S. Government Printing Office, 2009.